Marking-off techniques for metal fabrication

A BASIC COURSE

Cec Cox

Senior Head Teacher, Metal Fabrication
Sydney Technical College

Graham Meyer

Teacher, Metal Fabrication
Bankstown Technical College

Drawings by Graham Meyer

The McGraw-Hill Companies, Inc.

Beijing Bogotà Boston Burr Ridge IL Caracas
Dubuque IA Lisbon London Madison WI
Madrid Mexico City Milan Montreal New Delhi
New York San Francisco Santiago Seoul
Singapore St Louis Sydney Taipei Toronto

The **McGraw·Hill** Companies

Reprinted 1997, 2000 (twice), 2001, 2002, 2003, 2005, 2006, 2007 (thrice), 2008 , 2024

National Library of Australia Cataloguing-in-Publication data:

Cox, Cec.
   Marking-off techniques for metal fabrication.
   ISBN 10: 0 07 451004 5
   ISBN 13: 978 0 07 451004 9

   1. Metal-work. 2. Laying-out (Machine shop practice).
   3. Boilers. I. Meyer, Graham. II. Title.

671

Published in Australia by
McGraw-Hill Australia Pty Ltd
Level 33 World Square, 680 George Street Sydney, NSW 2000, Australia
Sponsoring editor: Stuart Lawrence
Copy editor: Daphne Rawling
Technical illustrators: Graham Meyer and Colin Bardill
Typesetters: Savage Type Pty Ltd
Printed in Australia by PEGASUS

Contents

v

12 Combined radial and parallel line development 257

13 Triangulation of transition pieces 277

14 Hoppers: Square or rectangular 325

Preface

This book has been written for students of metal fabrication in technical colleges and for those generally interested in this field. As teachers of boilermaking and metal fabrication, we have recognised the need for a text in marking-off methods. We feel this book meets that need.

The aim of the book is to present a practical approach to fabrication marking from basic principles to advanced layout work. Its special features include the use of calculations in layout work and an awareness of the need to control errors when marking. Also, the most economical method is used and allowance is made for thickness of material. The problems are explained in easy-to-follow steps. Explanations are placed alongside or opposite the graphic steps to avoid turning pages when working on a particular problem.

We would like to thank the following people and organisations for assistance in the preparation of this book: The Department of Technical and Further Education, NSW, for the use of items from the publication, Print Reading and Interpretation, and for the use of metal fabrication job sheets; McGraw-Hill Book Company Australia for assistance with writing and illustration, and permission to use extracts from Engineering Drawing by A. W. Boundy; the Amalgamated Metal Workers and Shipwrights Union for permission to use extracts from A Practical Treatise for Boilermakers by J. and H. Haddon; Maxitherm, NSW, for the use of photographs; and Harry Stevenson, retired teacher of boilermaking, for the use of some of his private notes.

Note

Do not measure from the diagrams except in exercises, because many figures are indication diagrams only and are not drawn to scale. All measurements on the diagrams are in millimetres unless stated otherwise.

CEC COX
GRAHAM MEYER

Geometric terminology for boilermaking

Terms used in geometry possess very exact meanings. The understanding of these meanings is important for the understanding of all future calculations, marking off and checking of jobs and layouts. Such concepts as a point, a line, etc., would normally need no explanation, but because each term is used in a strict sense in geometry, the following brief glossary is necessary.

Description of points and lines

A point	A line
Is a dimensionless geometric object having no property but location. Two methods of showing a point are shown below.	Is a thin continuous line or a series of points joined together.

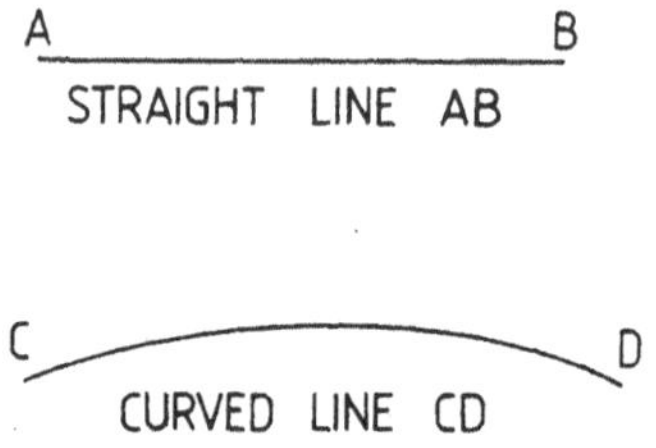

A.

POINT 'A'

B.
+

POINT 'B'

A ——————————— B

STRAIGHT   LINE   AB

C ⌢ D

CURVED   LINE   CD

A straight line	Parallel lines
Is the shortest distance between any two points, say A and B.	Are lines which run in the same direction. They are always an equal distance apart. They can never converge, cross or intersect.

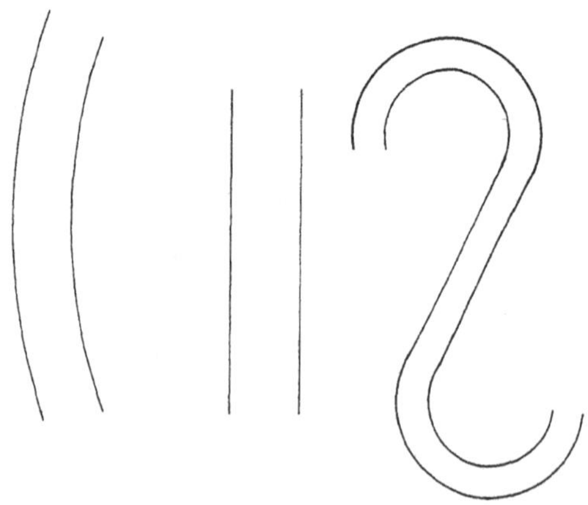

A ——————————— B

When two straight lines meet at a point they form an angle

A right angle	An acute angle
Is formed when two straight lines perpendicular to each other meet to form an angle.	Is an angle less than a right angle.

An obtuse angle	A reflex angle
Is an angle greater than a right angle but less than two right angles.	Is an angle greater than two right angles but less than four right angles.

A triangle is a figure bounded by three straight lines; the three angles so formed total 180°

Triangles are identified by:
1. their sides, 2. their angles.

An equilateral triangle	An isosceles triangle
Has all three sides equal in length and all angles also equal (60°).	Has two sides equal in length. Therefore the two base angles are also equal.
A scalene triangle	A right-angled triangle
Has all three sides different in length and three acute angles.	Is a triangle with one of the three angles 90°.

Other various shapes used in marking off

A square

Is a rectangle with all sides equal. The diagonal AC is equal in length to the diagonal *BD*.

A   B

D   C

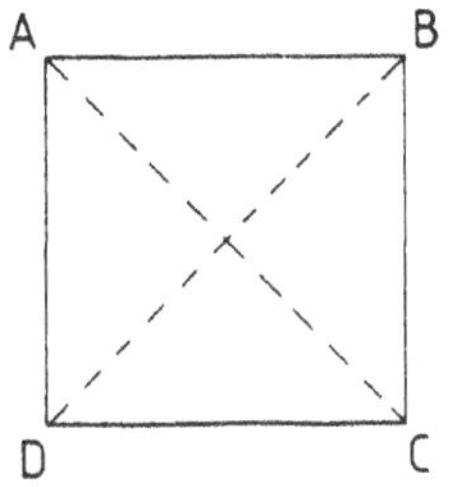

A rectangle

Is a four-sided figure with four right angles, and its opposite sides equal in length. The diagonal AC is equal in length to the diagonal *BD*.

A   B

D   C

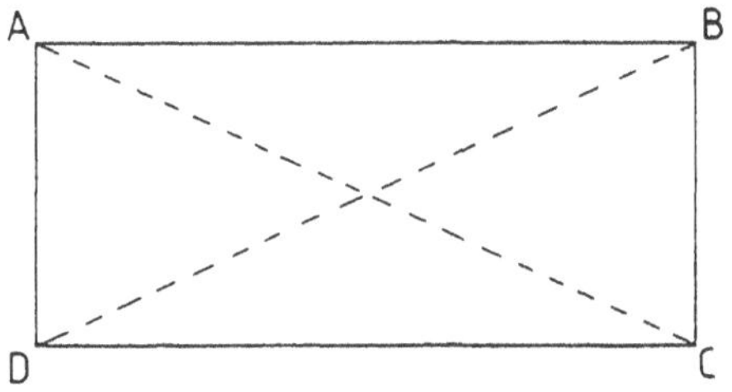

A parallelogram

Is a four-sided figure with the opposite sides parallel. It has two acute angles and two obtuse angles. *The diagonals are not equal.*

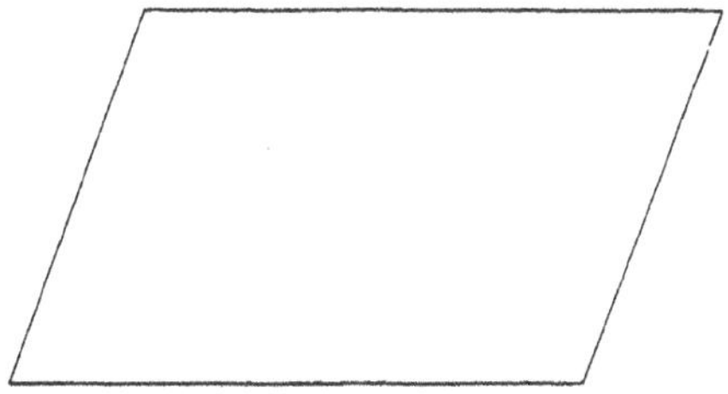

An obround

Is like an elongated hole. The ends are half circular in shape and the sides are parallel.

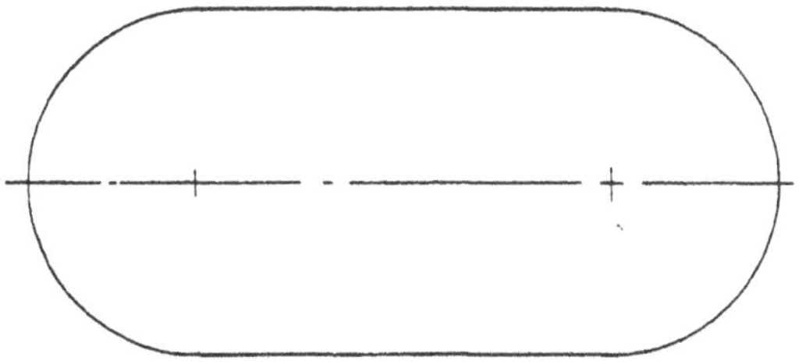

Other various shapes (cont'd)

A circle	An ellipse
Is a plane figure bounded by a curved line which lies throughout at an equal distance from a fixed point called the centre.	Is a plane curved figure with its major axis greater in length than its minor axis.
A cone	A cylinder
Is a solid figure which has a circular end, tapering to a point called the apex.	Is a solid figure with circular ends bounded by parallel sides.

Circles and parts of circles

It is most important to know and be able to recognise the various terms when using circles and part circles.

Centre	Circumference
This is the point around which the circle is scribed.	This is the length of the curved line that forms the circle.

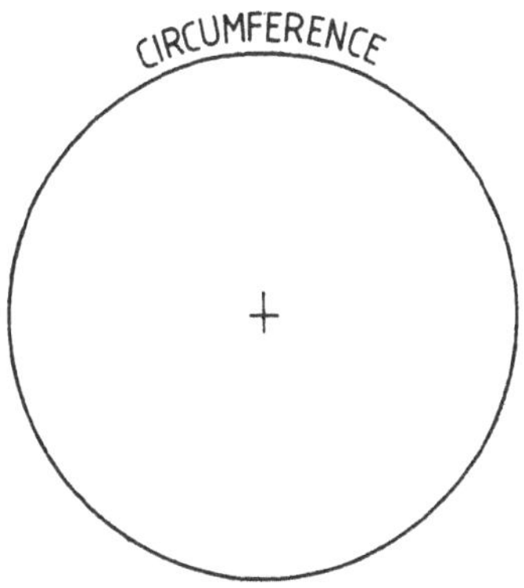

CENTRE POINT

CIRCUMFERENCE

Radius	Diameter
This is the distance from the centre of the circle to the circumference measured in a straight line.	This is double the length of the radius, or the distance from one side of the circumference to the other side, passing through the centre.

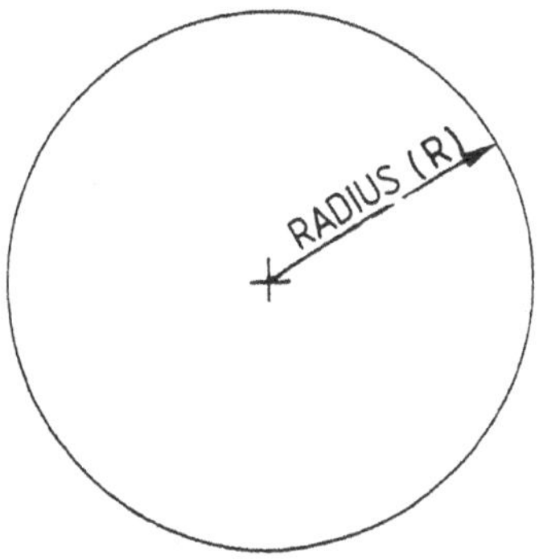

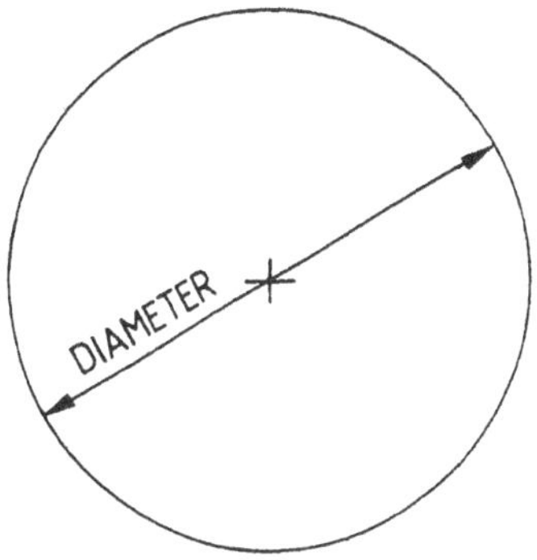

RADIUS (R)

DIAMETER

Circles and parts of circles (cont'd)

Concentric circles	Eccentric circles
Are circles having a common centre point but different radii.	Are circles which have no common centre, have different radii and are arranged so that one circle lies within the other.

An arc	A chord
Is any part of the circumference of any circle.	Is a line joining any two points on the circumference of any circle but not passing through the centre point.

Circles and parts of circles (cont'd)

A segment

Is the area bounded by an arc and its chord.

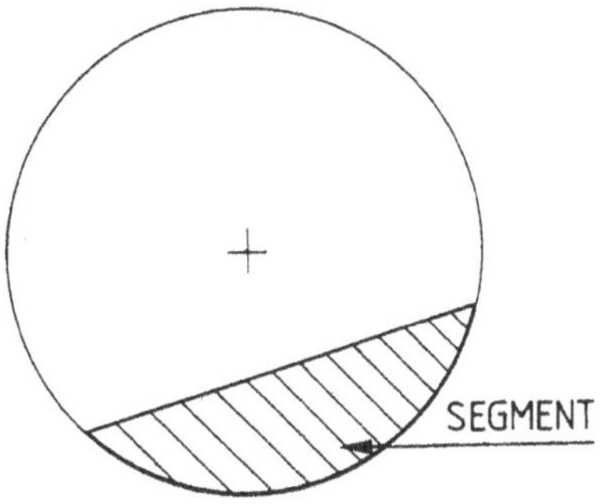

SEGMENT

A semicircle

Is an area exactly one-half of any given circle bounded by an arc.

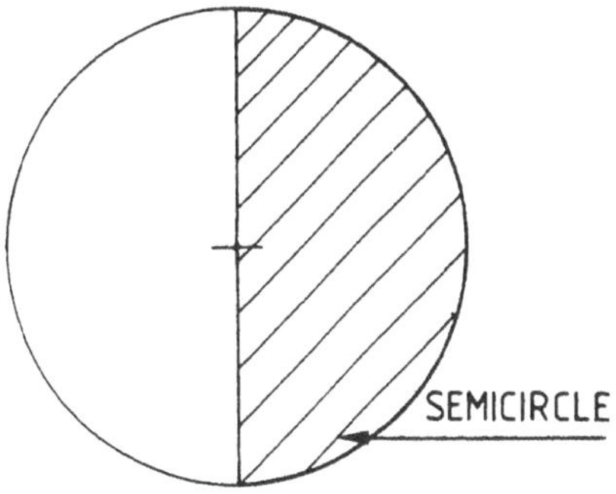

SEMICIRCLE

A sector

Is an area enclosed by two radii and an arc joining them.

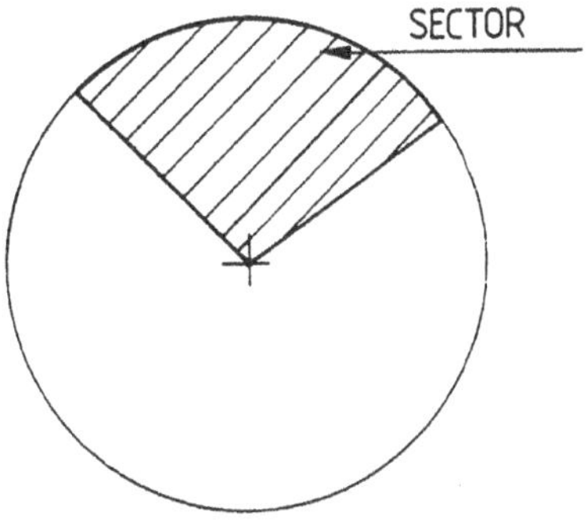

SECTOR

A quadrant

Is the area enclosed by two radii at right angles to each other, and the arc joining them. It forms exactly one-quarter of the circle.

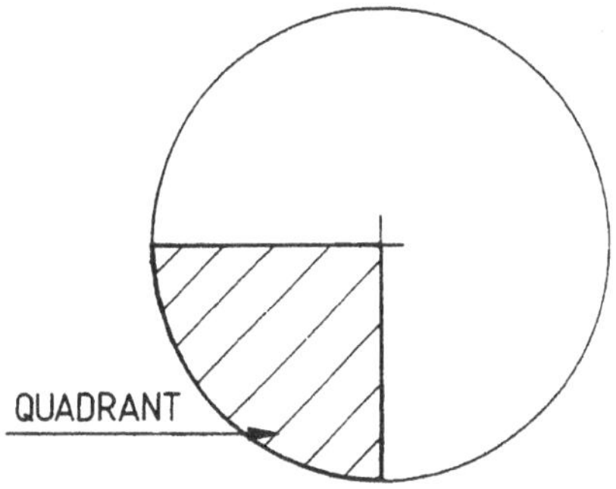

QUADRANT

Tangents

A tangent is a line, curve or surface, touching but not intersecting another line, curve or surface.

Line tangent to a circle

A line at 90° from straight line to centre of circle will locate the point of tangency.

Circles tangent to each other

A line connecting the two centres of the circles will locate the point of tangency.

Curves tangent to each other

A line connecting the two centres of the curves will locate the point of tangency.

Curve tangent to a straight line and circle

Locate the points of tangency as in the other three methods.

Bisecting

"Bisect" means to cut or divide into two equal parts.

Bisecting an angle	Bisecting a straight line
The bisector line *BD* is bisecting the right angle *ABC*.	The bisector line *AB* is bisecting the straight line *XY*.
Bisecting a curve	**Gusset plate item bisected by centre line**
The bisector line *CD* is bisecting the curve *AB*.	

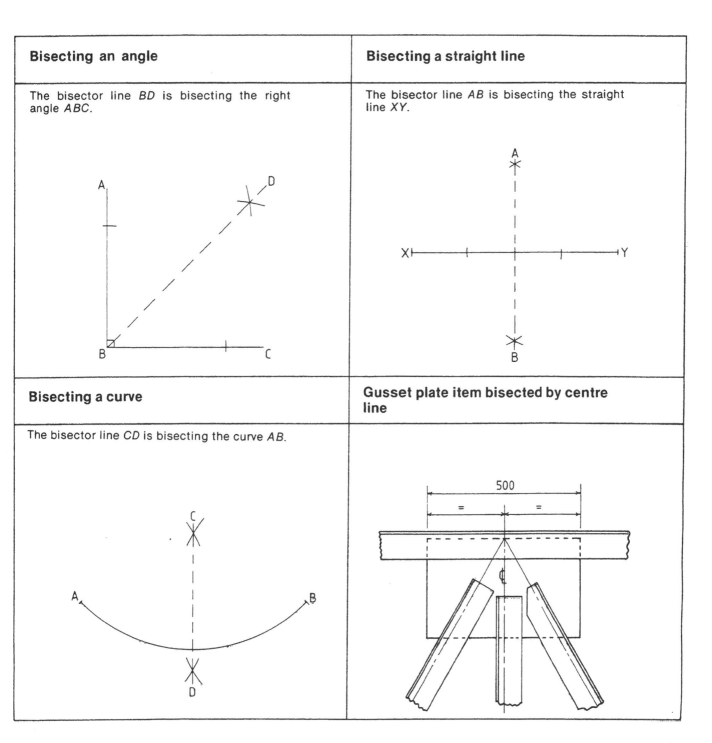

Other geometric terminology used in marking off

Camber	A hexagon
Is a surface deflected off straight in a continuous curve (a slightly arched surface) CAMBERED SURFACE	Is a six-sided figure.
A sphere	**A hemisphere**
Is a three-dimensional surface, all points of which are equidistant from a fixed centre point, e.g. a round ball.	Is half of a sphere.

Symmetry

Symmetry is the exact correspondence of form and constituent configuration on opposite sides of a dividing line or plane, or about a centre line or axis. It can be as a *mirror image* each side of the centre line (₵).

Example 1	Example 2
The half ellipse *ACB* is symmetrical about the major axis with the other half ellipse *ADB*.	The half ellipse *CAD* is symmetrical about the minor axis with the other half ellipse *CBD*.

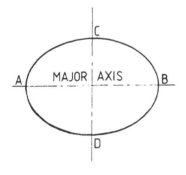 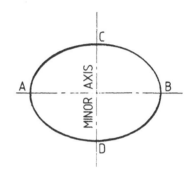

Below is a key diagram showing the use of symmetry on workshop drawings.

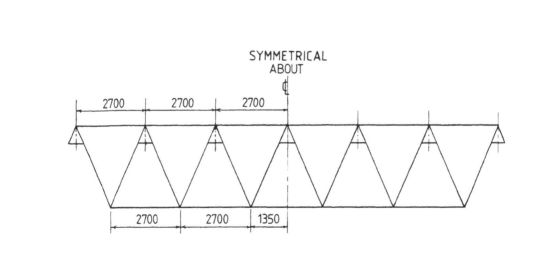

NOTES

2

Squaring of lines and marking lines parallel

It is most important for any marker-off to be able to geometrically construct square lines (right angles) and parallel lines. Most layouts are constructed around two square centre lines or from a square corner. If the marker-off is not accurate with these starting lines, then the rest of the layout will be out of square.

Geometric construction of square lines is preferred to using a plate square because of accumulative error. Geometric construction can be used on small and large work (using dividers, trammels or measuring tape) whereas the plate square can only be used on small work.

The following pages show step-by-step methods of squaring lines and marking lines parallel. The plate square is considered accurate only over its length and should be checked for square before use.

Problem

To construct a centre line at 90° to a base line.

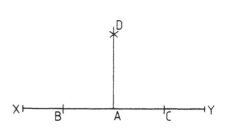

1	(a) Mark in base line *XY*. (b) Bisect base line *XY* at point *A*.

X⊢――――――――――⊣Y
;A

2	Using any radius and point *A* as centre, scribe arcs to locate points *B* and *C* on base line *XY*.

X⊢――B――――A――――C――⊣Y

3	Using any radius (larger than in step 2) and points *B* and *C* as centres, scribe arcs to intersect at point *D*.

D

X⊢――B――――A――――C――⊣Y

4	Join points *D* and *A*. The line *DA* is a centre line at 90° to the base line *XY*.

D

X⊢――B――――A――――C――⊣Y

Problem

To construct a vertical centre line at 90° to a horizontal centre line.

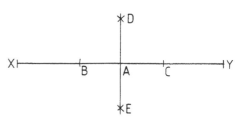

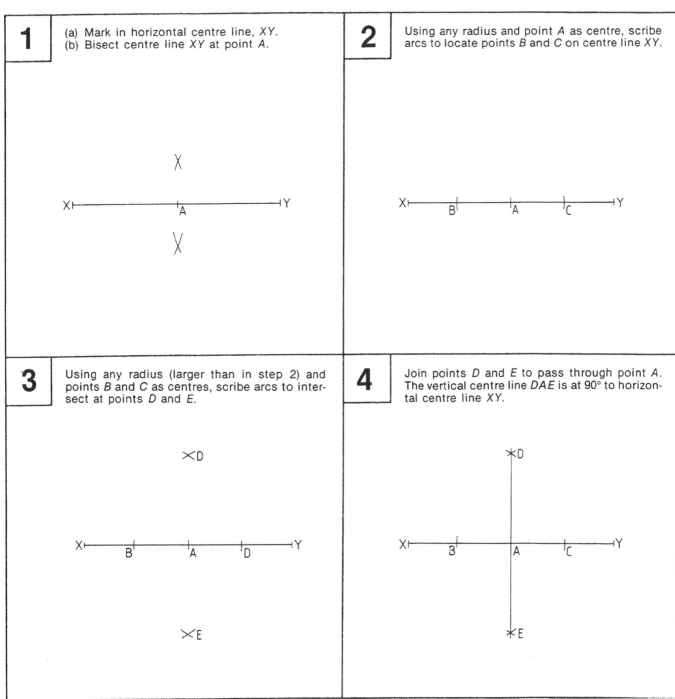

1

(a) Mark in horizontal centre line, *XY*.
(b) Bisect centre line *XY* at point *A*.

2

Using any radius and point *A* as centre, scribe arcs to locate points *B* and *C* on centre line *XY*.

3

Using any radius (larger than in step 2) and points *B* and *C* as centres, scribe arcs to intersect at points *D* and *E*.

4

Join points *D* and *E* to pass through point *A*. The vertical centre line *DAE* is at 90° to horizontal centre line *XY*.

Problem

To construct a line at 90° to a base line near the
end of the line using the 3-4-5 method.

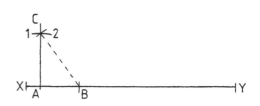

1	(a) Draw in base line *XY* and locate point *A* near one end of the line. (b) Using point *A* as centre and a distance equal to 3 units as radius, scribe arc to locate point *B* on base line *XY*.	2	Using point *A* as centre and a distance equal to 4 units as radius, scribe the arc 1-2.

1——2

3	Using point *B* as centre and a distance equal to 5 units as radius, scribe arc to locate point *C* on arc 1-2.	4	Join points *C* and *A* and the line *CA* is at 90° to the base line *XY*.

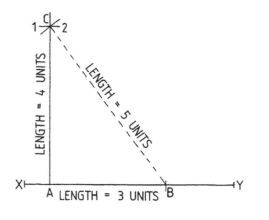

Problem

To construct a line at 90° to a base line near the end of the line using right-angled triangle in a semicircle method.

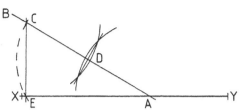

1	(a) Draw in base line *XY*. (b) Draw in line *AB* from any point and at any angle to base line *XY*.

2	(a) Mark in point *C* approximately above point *X* on line *BA*. (b) Bisect line *CA* at point *D*.

3	Using point *D* as centre and distance *CD* as radius, scribe arc to cut base line *XY* at point *E*.

4	Join points *C* and *E* and then the line *CE* is at 90° to the base line *XY*.

Problem

To draw a square line to a line from a point above that line.

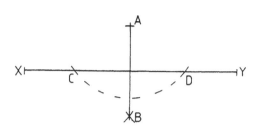

1	(a) Draw line *XY*. (b) Position point *A* above the line.

+ᴬ

X⊢————————————————⊣Y

2	Using point *A* as centre and any radius, scribe an arc to cut line *XY* at points *C* and *D*.

+ᴬ

X⊢————C————————D————⊣Y

3	Using points *C* and *D* as centres and same radius as in step 2, scribe arcs to intersect at point *B*.

+ᴬ

X⊢————C————————D————⊣Y

⤬B

4	Join points *A* and *B* and the line *AB* will be 90° to the line *XY*.

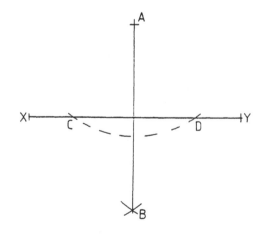

Problem

To construct a 90° line from the end of a base line using angle in a semicircle method.

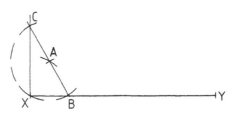

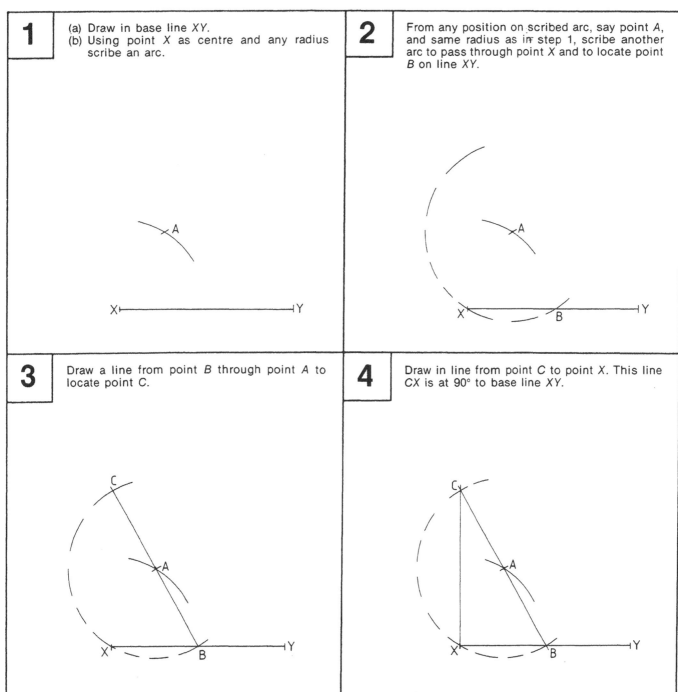

1
(a) Draw in base line *XY*.
(b) Using point *X* as centre and any radius scribe an arc.

2
From any position on scribed arc, say point *A*, and same radius as in step 1, scribe another arc to pass through point *X* and to locate point *B* on line *XY*.

3
Draw a line from point *B* through point *A* to locate point *C*.

4
Draw in line from point *C* to point *X*. This line *CX* is at 90° to base line *XY*.

Problem

To construct a 90° line from the end of a base line using another geometric method.

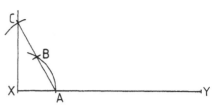

1 (a) Draw in base line *XY*. (b) Using point *X* as centre and any radius, scribe an arc to locate point *A*.	**2** (a) Using same radius as in step 1 and point *A* as centre, scribe an arc to locate point *B*. (b) Using same radius and point *B* as centre, scribe another arc.
3 Draw in line from point *A* through point *B* to locate point *C* on this third arc.	**4** Join points *C* and *X*. The line *CX* is at 90° to base line *XY*.

Exercises for squaring of lines

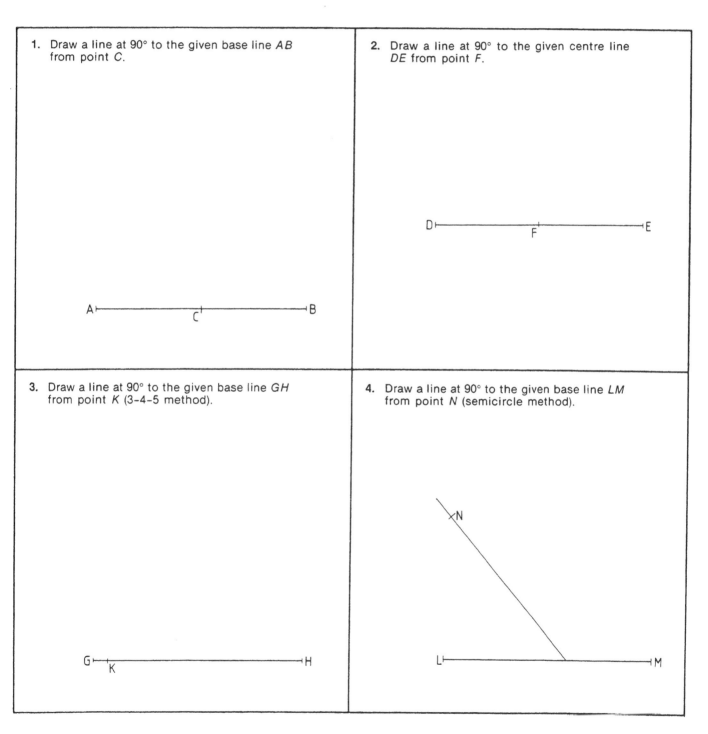

1. Draw a line at 90° to the given base line *AB* from point *C*.

2. Draw a line at 90° to the given centre line *DE* from point *F*.

3. Draw a line at 90° to the given base line *GH* from point *K* (3–4–5 method).

4. Draw a line at 90° to the given base line *LM* from point *N* (semicircle method).

Problem

To draw a straight line parallel to a given line at
a given distance apart.

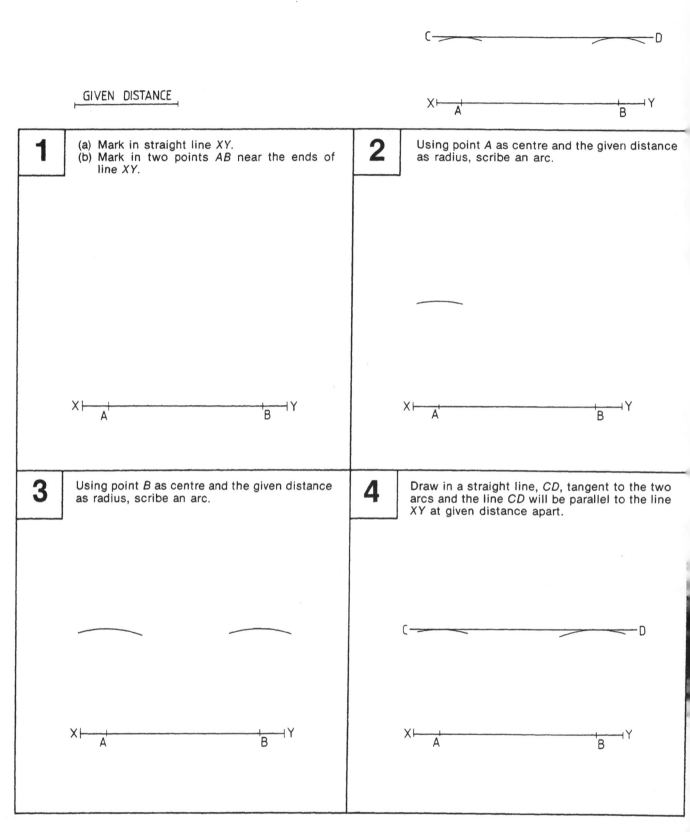

GIVEN DISTANCE

1
  (a) Mark in straight line *XY*.
  (b) Mark in two points *AB* near the ends of
      line *XY*.

2
Using point *A* as centre and the given distance
as radius, scribe an arc.

3
Using point *B* as centre and the given distance
as radius, scribe an arc.

4
Draw in a straight line, *CD*, tangent to the two
arcs and the line *CD* will be parallel to the line
XY at given distance apart.

Problem

To mark lines parallel to a given offset centre line, at a given distance each side of the centre line.

GIVEN DISTANCE

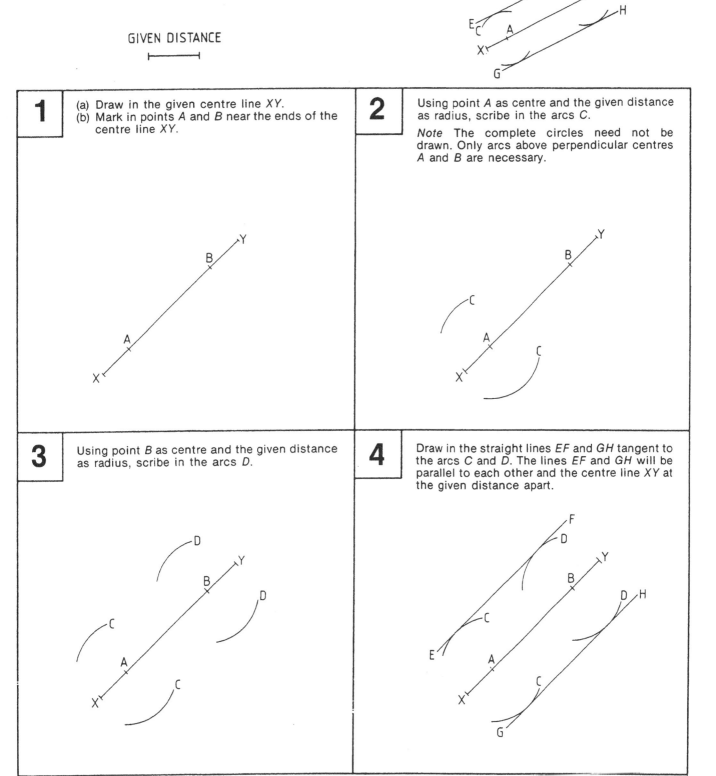

1
(a) Draw in the given centre line *XY*.
(b) Mark in points *A* and *B* near the ends of the centre line *XY*.

2
Using point *A* as centre and the given distance as radius, scribe in the arcs *C*.

Note The complete circles need not be drawn. Only arcs above perpendicular centres *A* and *B* are necessary.

3
Using point *B* as centre and the given distance as radius, scribe in the arcs *D*.

4
Draw in the straight lines *EF* and *GH* tangent to the arcs *C* and *D*. The lines *EF* and *GH* will be parallel to each other and the centre line *XY* at the given distance apart.

Problem

To draw a line parallel to an irregular curved line
at a given distance.

GIVEN DISTANCE

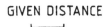

1	Line *AB* is the given irregular curved line.

B

A

2	From point *A*, mark off any number of equal distance points on line *AB*.

B

A

3	With the given distance as radius, scribe an arc from each of the located points on the line *AB*.

B

A

4	Draw a line tangent to all the scribed arcs and this line will be parallel to the given line *AB*.

B

A

Problem

To construct a 50 mm square from a given base line using dividers to square the end line and to position the parallel lines.

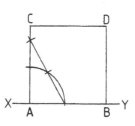

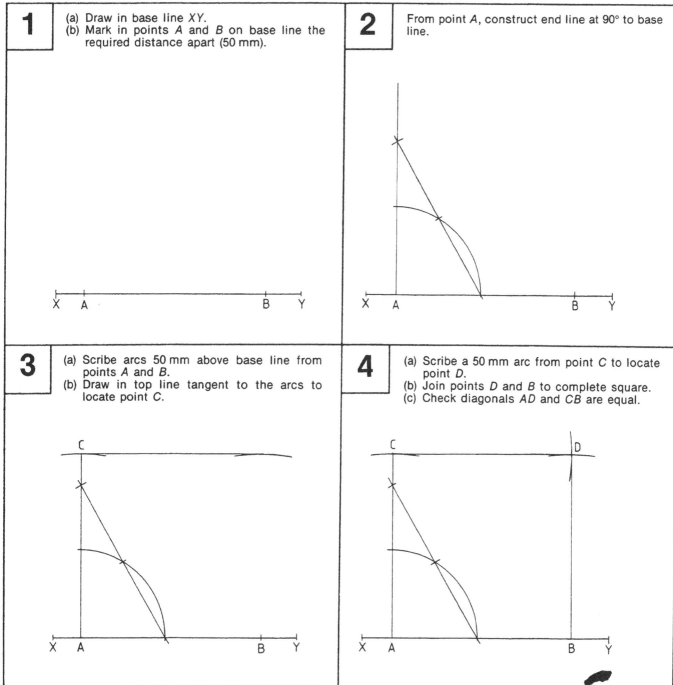

1
(a) Draw in base line *XY*.
(b) Mark in points *A* and *B* on base line the required distance apart (50 mm).

2
From point *A*, construct end line at 90° to base line.

3
(a) Scribe arcs 50 mm above base line from points *A* and *B*.
(b) Draw in top line tangent to the arcs to locate point *C*.

4
(a) Scribe a 50 mm arc from point *C* to locate point *D*.
(b) Join points *D* and *B* to complete square.
(c) Check diagonals *AD* and *CB* are equal.

Problem

To draw parallel development lines on a side view of a cylinder from half-end views.

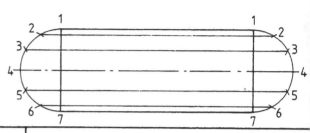

1	Draw in side view of cylinder.

2	Draw in half-end view of top end and divide into, say, six equal spaces and number 1 to 7.

3	Draw in half-end view of other end, divide into six equal spaces and number 1 to 7.

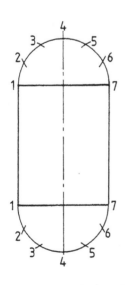

4	Join the corresponding numbered points to locate parallel development lines.

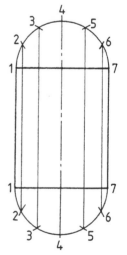

Exercises for drawing perpendicular and parallel lines

1. Draw a line at 90° to base line *XY*, from point *X*.

X⊢————————————⊣Y

2. Using another method than in exercise 1, draw a line at 90° to base line *XY*, from point *X*.

X⊢————————————⊣Y

3. Draw in lines parallel to centre line, from given points on half-end view of cylinder (development lines).

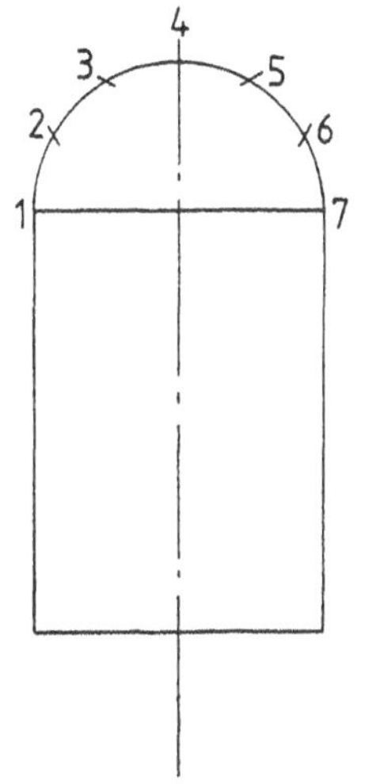

4. Construct a 65 mm square from given base line *XY*, using one square end line plus parallel lines.

X⊢————————————⊣Y

Exercises (cont'd)

5. Draw a line at 90° to the given base line *AB* from point *C* above the base line.

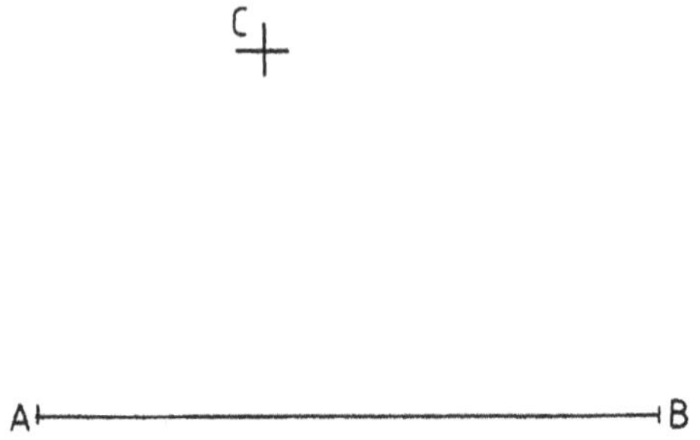

C
+

A ├─────────────────────────────┤ B

6. Draw a line parallel to base line *DE* at the given distance of 45 mm.

D ├─────────────────────────────┤ E

7. Draw lines parallel to centre line *FG* at given distance of 20 mm each side.

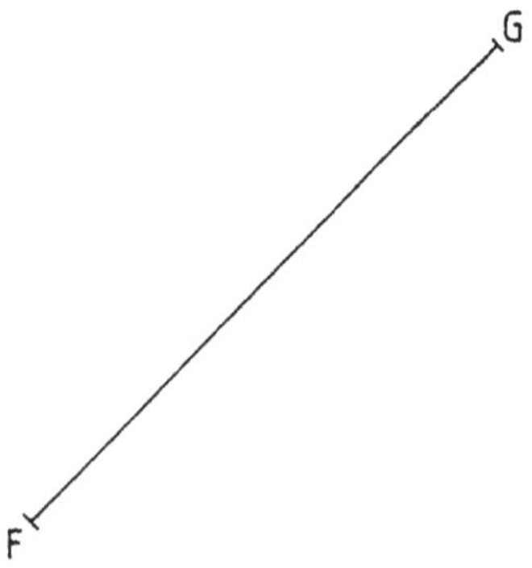

G

F

8. Draw a line parallel to given curved line *HJ* at given distance 20 mm.

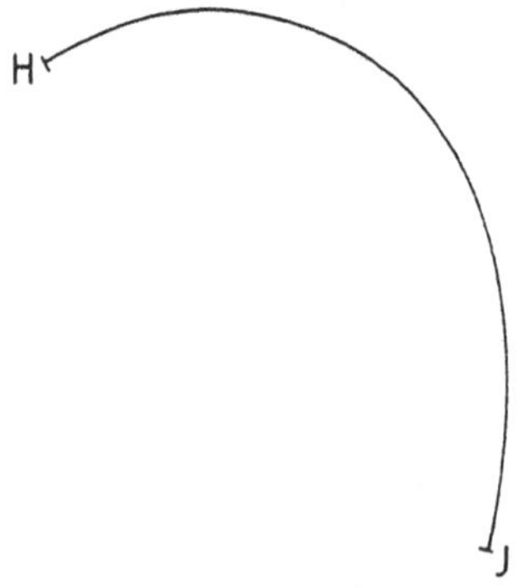

H

J

3

Dividing lines, circles and angles

Knowledge of geometric bisection of lines and angles, coupled with the division of circles into part circles, forms an integral part of a marker-off's range of skills. After completing this chapter with the practice examples, the following chapters will be easier to interpret and complete.

These examples are drawn on only a small scale using dividers, but the same principles are followed on larger work using trammels or measuring tape.

Note
Do not measure from the diagrams except in exercises, because many figures are indication diagrams only and are not drawn to scale. All measurements on the diagrams are in millimetres unless stated otherwise.

Problem

To bisect a given straight line (divide into two
equal sections).

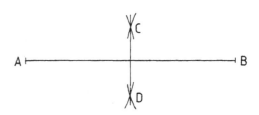

1	(a) Let *AB* be the given straight line. (b) Using point *A* as centre and the radius slightly larger than half the length of *AB*, scribe an arc.

2	Using point *B* as centre and same radius as in step 1, scribe an arc to intersect the first arc to locate points *C* and *D*. *Note* The complete arcs need not be drawn. Only the intersecting points *C* and *D* are necessary, especially if drawing the arcs in full confuses the marking.

3	Join points *C* and *D* with a straight line and the given line *AB* will be bisected by line *CD*.

Problem

To divide a given straight line into four equal spaces.

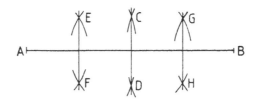

1	(a) Let *AB* be the given straight line. (b) Let *CD* bisect the given straight line at point *X*. (c) Using point *X* as centre and any radius larger than half the length *AX*, scribe arcs each side of point *X*.

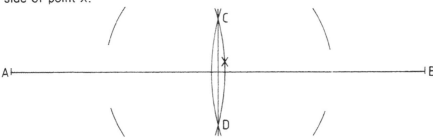

2	Using points *A* and *B* as centres and same radius as in step 1, scribe arcs to intersect the arcs from step 1 to locate points *E*, *F*, *G* and *H*.

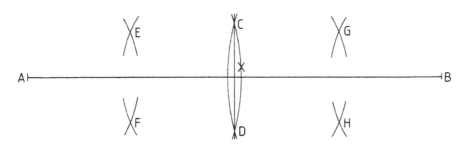

3	(a) Join point *E* to *F* and point *G* to *H*. The line *AB* is divided into four equal spaces. (b) Continue to bisect the equal spaces to divide the line into eight, sixteen and thirty-two equal spaces.

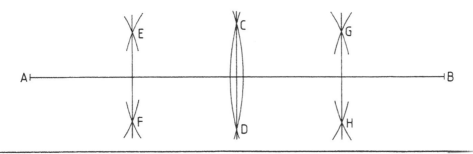

Problem

To divide the given straight line into six equal spaces.

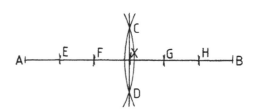

1

(a) Let *AB* be the given straight line 115 mm long.

(b) Let *CD* bisect the given line *AB* at point *X*.

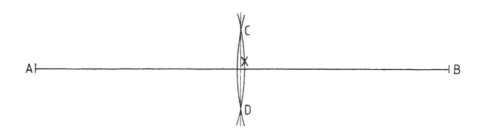

2

(a) Measure the distance *AX* and mathematically divide it by 3. This will give you the correct radius to use (57.5÷3 = 19.17 mm).

(b) Using the radius from step 2(a), scribe arcs to locate points *E* and *F* between points *A* and *X*, then double check from point *X* back to point *A*.

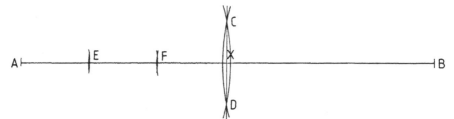

3

(a) Using the same radius as in step 2, locate points *G* and *H* between points *X* and *B*, and double check from point *B* back to point *X*.

(b) The line *AB* is then divided into six equal spaces by points *E*, *F*, *X*, *G* and *H*.

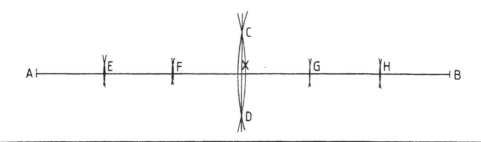

Problem

To divide a given straight line into twelve equal spaces.

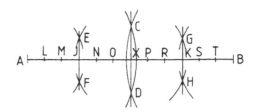

1   Let *AB* be the given straight line 115 mm long divided into four equal spaces by points *J*, *X* and *K*.

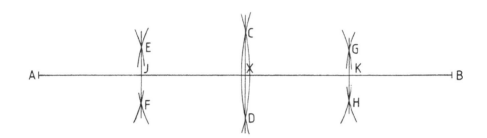

2   (a) Measure distance *AJ* and mathematically divide this by 3. This will give you the correct radius to use (28.75 ÷ 3 = 9.58 mm).

       (b) Using radius from step 2(a), locate points *L* and *M* between points *A* and *J*, starting from point *A* to *J* and back checking from point *J* to *A*.

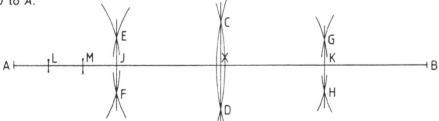

3   (a) Using the same radius as in step 2(b) and the same procedure, locate points *N* and *O* between points *J* and *X*, points *P* and *R* between *X* and *K*, then points *S* and *T* between points *K* and *B*.

       (b) The given straight line is then divided into twelve equal spaces.

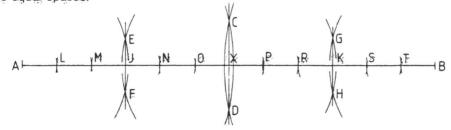

Practical example

This is a drawing of an actual job to demonstrate the need for dividing given lines into a set number of equal spaces.

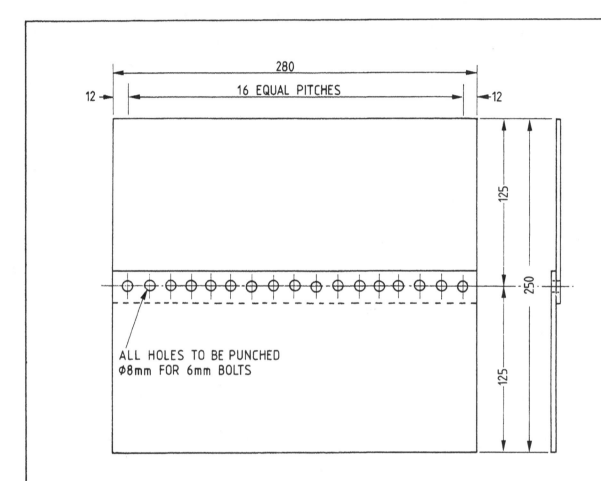

Workshop instructions. Job No. 2, Stage 1. Lap joint.

1. Each student to mark off, on his layout plate, one of the two plates.
2. After the layout has been checked by the teacher, the student can mark off one plate on the actual job material — include production instructions.
3. Each student to complete one plate only and then "pair up" to form the lap joint.

Associated exercise

Lay out one plate full size, showing all construction lines, dimensions, calculations and operation sequence.

Material

Two off 300 mm × 150 mm × 3 mm black mild steel plate (FL).

This sheet is an example of what is required for associated exercise for Job 2, Stage 1. Lap joint.

Operational sequence

1. Construct rectangle 280 mm × 137 mm using base line and centre line method.
2. Check rectangle for square by measuring the diagonals.
3. Calculate the edge landing.
   1½ × hole diam. = 1½ × 8 = 12 mm
4. Mark in centre line of holes 12 mm in parallel from top edge.
5. Mark in end holes 12 mm from each end. This now gives you three holes marked by using the centre line of the plate as the third hole marked.
6. Bisect the distances between the three hole centres. There are now five hole centres.
7. Bisect the distances between the five hole centres. There are now nine hole centres.
8. Bisect the distances between the nine hole centres. There are now seventeen hole centres.
9. Because the sixteen pitches are along a straight line they give you seventeen holes.
10. Shear all the edges.
11. Punch all the holes.
12. Remove all burrs from holes and edges, bolt together and present for marking.

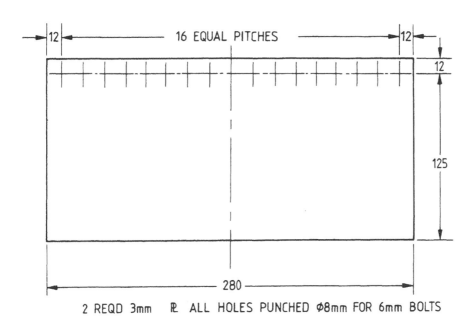

2 REQD 3mm ℞ ALL HOLES PUNCHED ⌀8mm FOR 6mm BOLTS

Exercises for division of lines

1. Bisect line *XY* of given length into two equal parts.

X├───────────────────────────────────────┤Y

2. Divide line *AB* of given length into four equal parts.

A├───────────────────────────────────────┤B

3. Divide line *CD* of given length into six equal parts.

C├─────────────────────────────────┤D

4. Divide line *EF* of given length into eight
equal parts.

E├────────────────────────────────────┤F

5. Divide line *GH* of given length into twelve
equal parts.

G├────────────────────────────────┤H

6. Divide line *JK* of given length into sixteen
equal parts.

J├────────────────────────────────────┤K

Problem

To bisect any given angle (acute, square, obtuse and reflex).

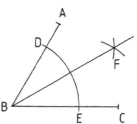

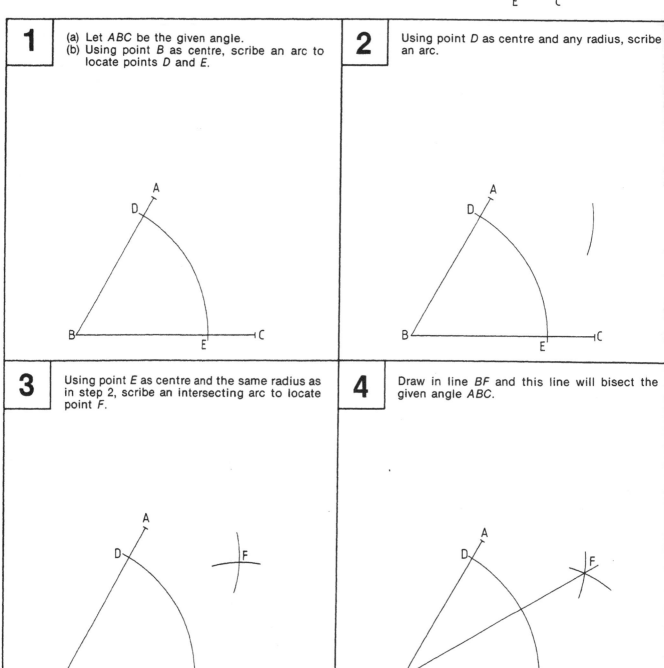

| **1** | (a) Let *ABC* be the given angle.
(b) Using point *B* as centre, scribe an arc to locate points *D* and *E*. |

| **2** | Using point *D* as centre and any radius, scribe an arc. |

| **3** | Using point *E* as centre and the same radius as in step 2, scribe an intersecting arc to locate point *F*. |

| **4** | Draw in line *BF* and this line will bisect the given angle *ABC*. |

Problem

To bisect any given curve.

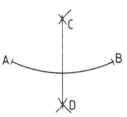

1	Draw in given curve *AB*.	**2**	Using point *A* as centre and any radius larger than half of curve *AB*, scribe arcs both sides of curve *AB*.

3	Using point *B* as centre and same radius as in step 2, scribe arcs both sides of curve *AB* to intersect first arcs to locate points *C* and *D*.	**4**	Join points *C* and *D* and the curve *AB* is then bisected by line *CD*.

Problem

To find the centre point of any given circle using geometric method.

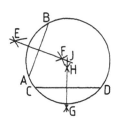

1 Draw chords *AB* and *CD* anywhere on the given circle.

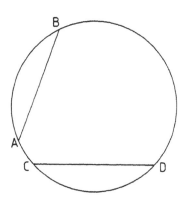

2 Bisect chord *AB* by using *A* and *B* as centres to locate points *E* and *F*.

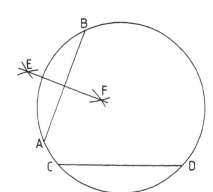

3 Bisect chord *CD* using *C* and *D* as centres to locate points *G* and *H*.

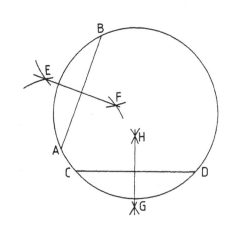

4 Extend lines *EF* and *GH* to intersect and the intersection point *J* will be the centre of the circle.

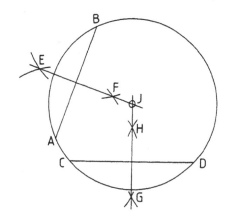

Problem

To find the centre point of any given circle using alternative method.

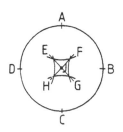

1	Mark four approximately equal spaces on given circle to locate the points A, B, C and D.	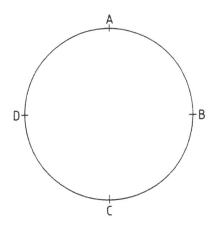

2 Using points A and C as centres and any radius smaller than radius of the circle, scribe two arcs.

3 Using points B and D as centres and same radius as in step 2, scribe two arcs to intersect the other two arcs, locating points E, F, G and H.

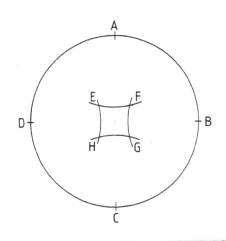

4 Join point E to G and point F to H. The centre point of the circle is where these two lines intersect at point J.

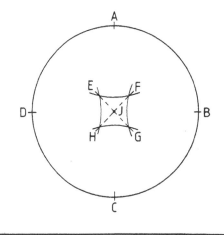

Problem

(a) To divide a given circle into two equal sections.

(b) To divide a given circle into four equal sections.

(a)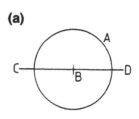

(b)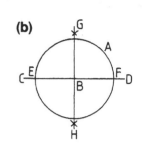

1
(a) Let *A* represent the given circle drawn around the centre point *B*.

(b) The given circle *A* will be divided into two equal sections by the straight line *CD* drawn through the centre point *B*.

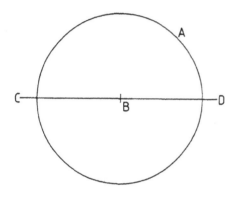

2
Circle *A* intersects centre line *CD* at points *E* and *F*.

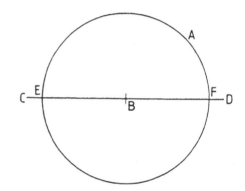

3
Using points *E* and *F* as centres and a radius larger than the radius of the given circle, scribe arcs both sides of the centre line *CD* to locate points *G* and *H*.

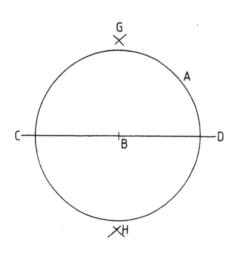

4
Join points *G* and *H* passing through centre point *B* and the lines *GH* and *CD* divide the circle into four equal sections.

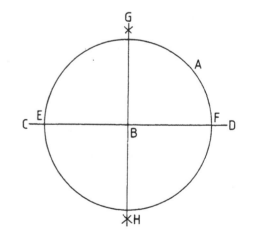

Problem

To divide a given circle into six equal sections.

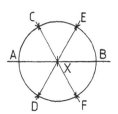

1	(a) Draw the given circle around the centre point X. (b) Halve the given circle with the centre line AB.	**2**	Using point A as centre and the radius of the given circle as radius, scribe points C and D on the given circle.

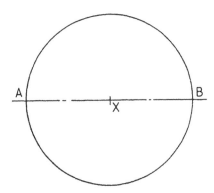

3	(a) Using point B as centre and same radius as in step 2, scribe points E and F on the given circle. (b) Double check that the distances CE and DF are equal to the radius used in steps 2 and 3(a).	**4**	Join points C to F and D to E through centre point X. The given circle is now divided into six equal sections by lines CF, ED, and AB.

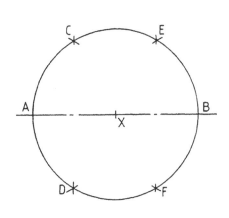

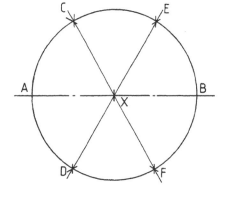

45

Problem

To divide a given circle into eight equal sections.

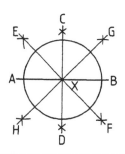

1 Draw given circle around centre point X and divide the given circle into four equal sections.	**2** (a) Using points A and C as centres and any radius, scribe intersecting arcs to locate point E. (b) Using points B and D as centres and any radius, scribe intersecting arcs to locate point F.

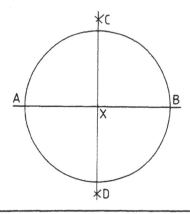

3 (a) Using points C and B as centres and any radius, scribe intersecting arcs to locate point G. (b) Using points A and D as centres and any radius, scribe intersecting arcs to locate point H.	**4** (a) Join points E and F through centre point X. (b) Join points G and H through centre point X. (c) The given circle is then divided into eight equal sections by lines AB, CD, EF and GH.

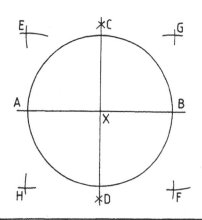
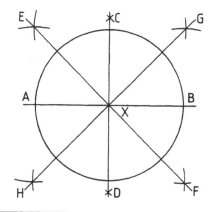

Problem

To divide a given circle into twelve equal sections.

---

1 Draw the given circle around the centre point *X* and divide the circle into four equal sections.

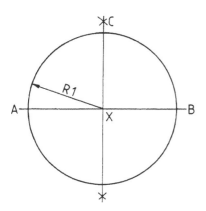

2 (a) Using point *C* as centre and radius *R1*, scribe arcs to locate points *E* and *F* on the given circle.
(b) Using point *D* as centre and radius *R1*, scribe arcs to locate points *G* and *H* on the given circle.

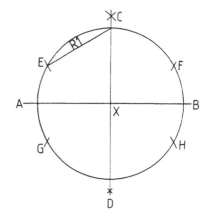

3 (a) Using point *A* as centre and radius *R1*, scribe arcs to locate points *J* and *K* on the given circle.
(b) Using point *B* as centre and radius *R1*, scribe arcs to locate points *L* and *M* on the circle.

4 (a) Join points *E* and *H* through centre *X*.
(b) Join points *J* and *M* through centre *X*.
(c) Join points *L* and *K* through centre *X*.
(d) Join points *F* and *G* through centre *X*.
The given circle is now divided into twelve equal sections.

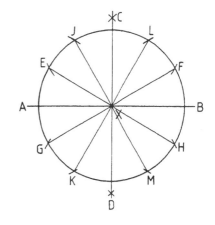

Practical examples

The following are taken from actual drawings.
They show the need for division of circles.
1. Pitching of holes on side plate of fan casing
2. Pitching of holes on connecting flange
3. Pitching of holes on end flange of fan casing
4. Pitching of holes on angle iron flange
5. Dividing of circular end in development by triangulation and parallel line

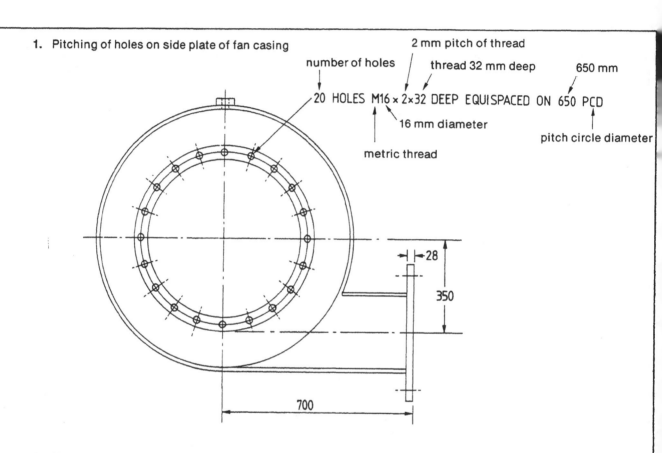

1. Pitching of holes on side plate of fan casing

number of holes

2 mm pitch of thread

thread 32 mm deep

650 mm

20 HOLES M16 × 2×32 DEEP EQUISPACED ON 650 PCD

16 mm diameter

metric thread

pitch circle diameter

28

350

700

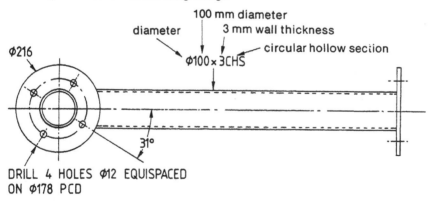

2. Piching of holes on connecting flange

100 mm diameter

3 mm wall thickness

diameter

circular hollow section

Ø216

Ø100 × 3CHS

31°

DRILL 4 HOLES Ø12 EQUISPACED
ON Ø178 PCD

3. Pitching of holes on end flange of fan casing

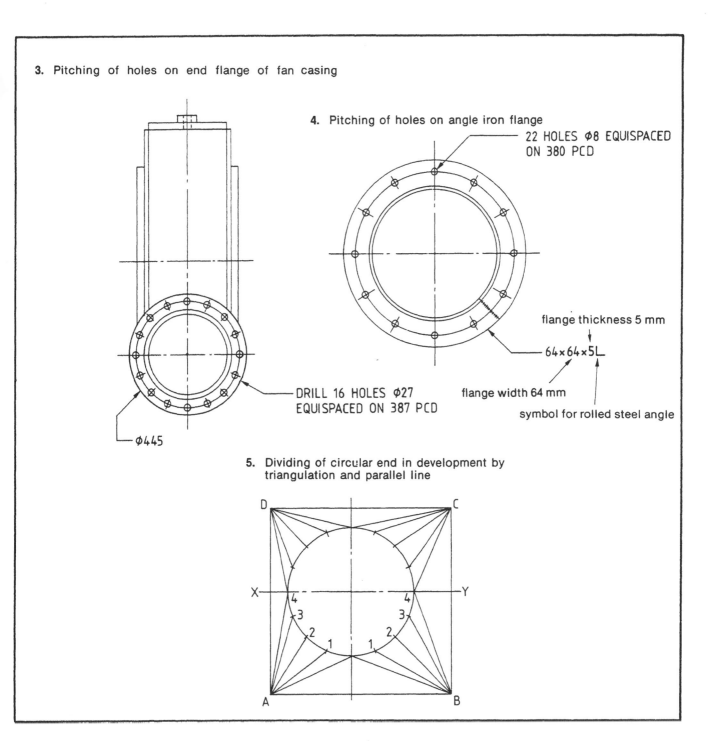

4. Pitching of holes on angle iron flange

22 HOLES φ8 EQUISPACED ON 380 PCD

flange thickness 5 mm

64×64×5L

flange width 64 mm

symbol for rolled steel angle

DRILL 16 HOLES φ27 EQUISPACED ON 387 PCD

φ445

5. Dividing of circular end in development by triangulation and parallel line

49

Exercises for bisecting angles and dividing circles

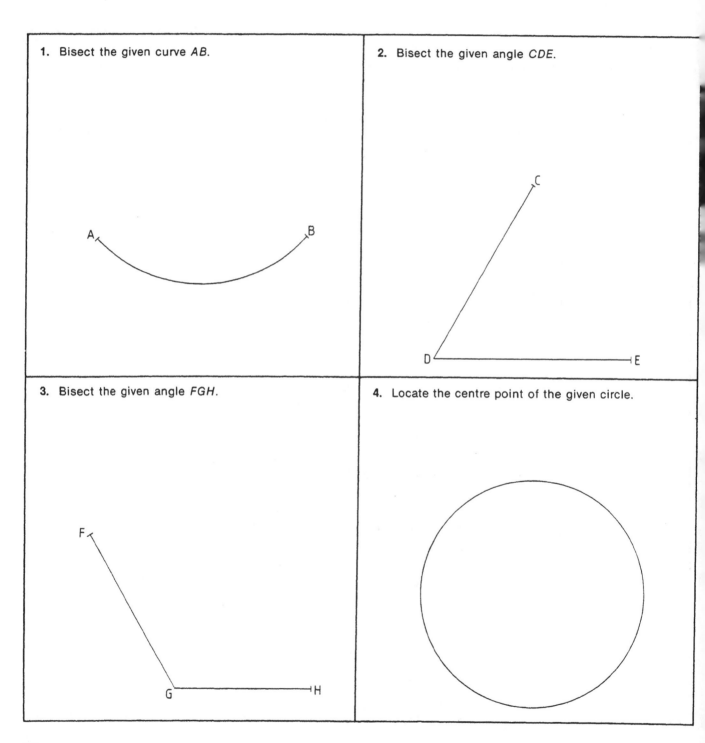

1. Bisect the given curve *AB*.

2. Bisect the given angle *CDE*.

3. Bisect the given angle *FGH*.

4. Locate the centre point of the given circle.

Exercises (cont'd)

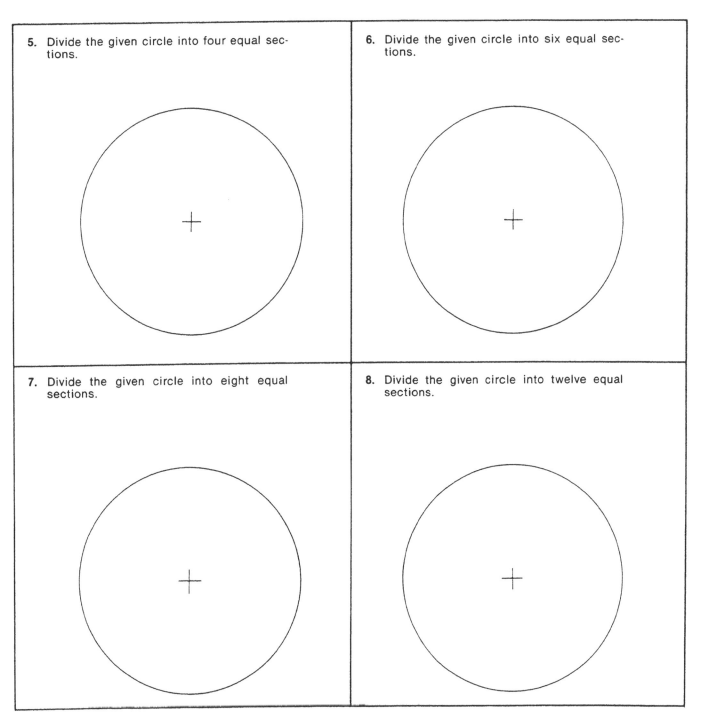

5. Divide the given circle into four equal sections.

6. Divide the given circle into six equal sections.

7. Divide the given circle into eight equal sections.

8. Divide the given circle into twelve equal sections.

NOTES

Constructing angles and transferring angles

Being competent in geometric construction of any angle at a given degree is vital to any boilermaker who desires to be a marker-off.

Some angles are of an unusual degree for geometric construction so we are supplying you with an alternative method, combining geometric construction with mathematics.

Transferring of angles from layout to actual job is also most important for eliminating welding gaps and hole misalignment.

Note
Do not measure from the diagrams except in exercises, because many figures are indication diagrams only and are not drawn to scale. All measurements on the diagrams are in millimetres unless stated otherwise.

Using bisection and trisection, most angles can be constructed from the following angles.

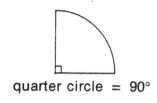

quarter circle = 90°

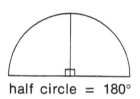

half circle = 180°

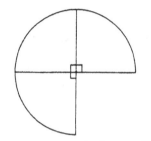

three-quarter circle = 270°

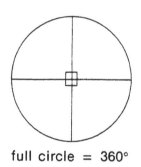

full circle = 360°

The radius of any circle, stepped around the circumference of that circle as chords, will give you six equal spaces. Now a circle = 360°, so each of these segments will have an angle of 360° ÷ 6 = 60°.

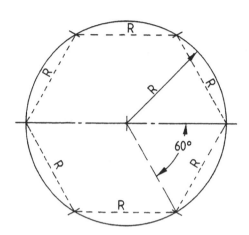

The following pages give examples of constructing and transferring angles using step-by-step method for clarity. These are followed with some examples for you to practise on.

We dealt with constructing a 90° angle in Chapter 2, so many of these exercises will start from the knowledge of constructing a 90° angle.

After mastering the basic 90°, 60°, 45°, 30° and 20°, the other common angles are constructed mainly by bisecting and/or trisecting the basic angles.

The latter part of this chapter deals with constructing the uncommon angles using geometric construction and maths.

Problem

To construct an angle of 45°.

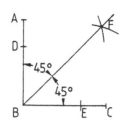

1	(a) Let angle *ABC* be equal to 90°. (b) Using any radius and point *B* as centre, scribe arcs to locate points *D* and *E*.

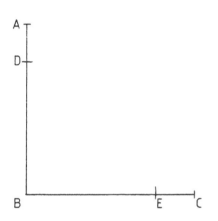

2	Using any radius and point *D* as centre, scribe an arc.

3	Using same radius as in step 2 and point *E* as centre, scribe an intersecting arc to locate point *F*.

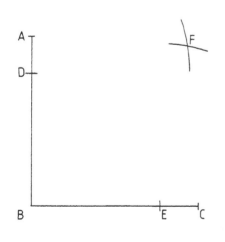

4	Join point *F* to point *B* and the line *BF* bisects the 90° angle to produce two angles of 45°.

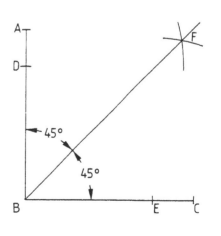

Problem

To construct an angle of 60°.

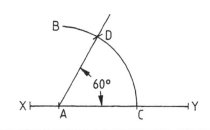

1	(a) Let *XY* be a given straight line. (b) Place point *A* in any position on the line.

2	Using any radius and point *A* as centre, scribe the arc *BC*.

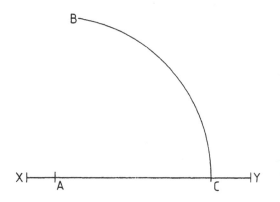

3	Using same radius as in step 2 and point *C* as centre, scribe an arc to locate point *D*.

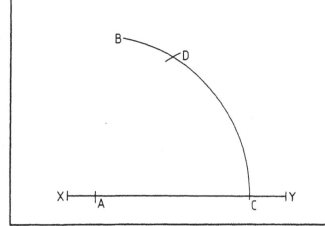

4	Join points *D* and *A* and the angle *DAC* will be 60°.

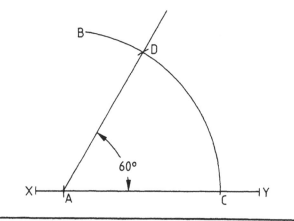

Problem

To construct a 30° angle from a 90° angle.

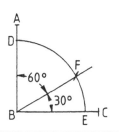

---

1 Let the lines *AB* and *BC* enclose a given 90° angle.

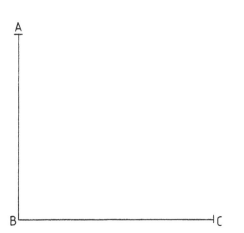

2 Using any radius and point *B* as centre, scribe the arc *DE* to cut lines *AB* and *BC*.

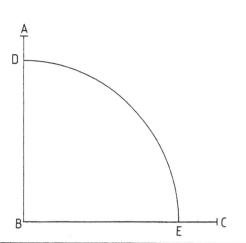

3 Using same radius as in step 2 and point *D* as centre, scribe an arc to cut the arc *DE* at point *F*.

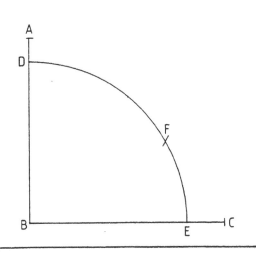

4 Join points *F* and *B* and the angle *FBE* will be 30°. The remaining angle *DBF* will be 60°.

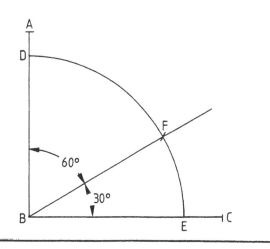

Problem

To construct a 30° angle from a straight line and a 60° angle.

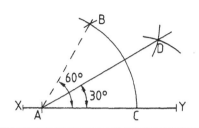

1

(a) Let *XY* be the given straight line.
(b) Let the angle *BAC* be the given 60° angle.

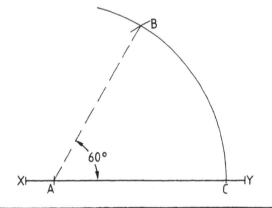

2

Using point *C* as centre and any radius, scribe an arc.

3

Using point *B* as centre and the same radius as in step 2, scribe an arc to locate point *D*.

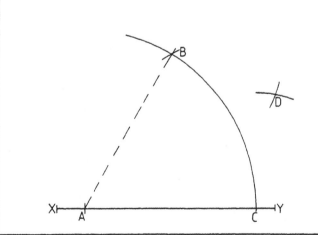

4

Join points *A* and *D* and the angle *DAC* will be 30°.

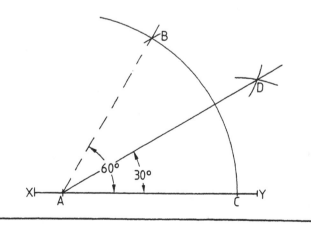

Problem

To construct an angle of 20°.

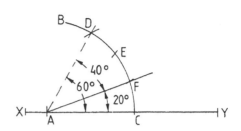

1	The angle *DAC* is a constructed 60° angle.

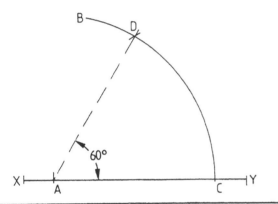

2	Using a radius equal to one-third the distance around the curve *DC* and point *D* as centre, step off three equal spaces to point *C*. (Use trial and error method.)

3	Using same radius as in step 2 and starting at point *C*, back check the three equal spaces to point *D* to accurately locate points *E* and *F*.

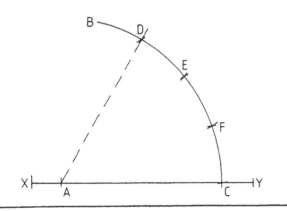

4	Join points *F* and *A* and the angle *FAC* will be 20° leaving the remaining angle to equal 40°.

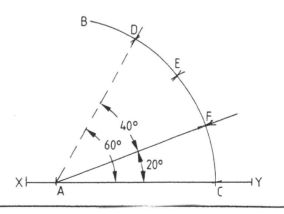

Using maths and geometric construction, any angle can be formed. The following formula will give you the distance around any given curve or circle ($1° = \frac{1}{360}$ of a circle):

$$1° = \frac{d \times \pi}{360} \quad \text{or} \quad 1° = \frac{(r \times 2) \times \pi}{360}$$

where d = diameter of circle,
π = 3.1416,
360 = no. of degrees in a circle,
r = radius.

Example

Construct an angle of 128° on a 100 mm diam. circle.

$$8° = \left(\frac{d \times \pi}{360}\right) \times 8$$

$$= \left(\frac{100 \times 3.1416}{360}\right) \times 8$$

$$= \left(\frac{314.16}{360}\right) \times 8$$

$$= 0.872\,66 \times 8$$

$$= 6.981\,28 \text{ mm}$$

There is an alternative method using geometric construction and maths to form an uncommon angle.

114.6 mm diam. circle or 57.3 radius circle has a circumference of 360.027 36 mm.

In this circle, the arc of 1° = 1 mm.

$$\therefore 1° = \frac{d \times \pi}{360}$$

$$= \frac{114.6 \times 3.1416}{360}$$

$$= \frac{360.027\,36}{360}$$

$$= 1.000\,076 \text{ mm}$$
$$\text{say, 1 mm}$$

So 1 mm measured around the circumference of a circle of 114.6 diam. = 1°.

Example

Construct a 69° angle on a 57.3 mm radius.

$\therefore$ Construct 60° angle geometrically then add 9 mm to equal 69°.

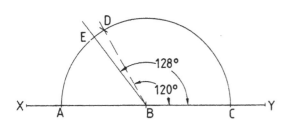

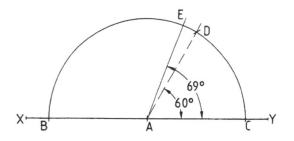

1. Let *DBC* be a geometrically constructed 120° angle.
2. From point *D* measure 6.98 mm around the semicircle towards point *A* to locate point *E*.
3. Join point *E* to point *B*.
4. The angle *EBC* will be 128°.

1. Let *DAC* be a 60° angle on a 57.3 mm radius semicircle.
2. From point *D*, measure around the curve 9 mm to locate point *E*.
3. Join points *E* and *A*.
4. The angle *EAC* will be 69°.

The previous alternative method is worked on a small scale, but if you want to work on a larger scale the same principle can apply.

In a 114.6 mm diam. circle, 1° = 1 mm around the circumference.

∴ In a 1146 mm diam. circle, 1° = 10 mm around the circumference

∴ In a 11 460 mm diam. circle, 1° = 100 mm around the circumference

and so on in multiples of 10.

Also, 114.6 mm × 2 = 229.2 mm.

∴ In a 229.2 mm diam. circle, 1° = 2 mm around the circumference

∴ In a 343.8 mm diam. circle, 1° = 3 mm around the circumference

and so on, by multiplying 114.6 × 4, 5, 6, 7, 8, 9, to suit whatever size you want to work in.

By the above tables:

330 mm measured around a 1146 mm diam. circle will give an angle of 33° (33 × 10)

26 mm measured around a 114.6 mm diam. circle will give an angle of 26° (26 × 1)

2600 mm measured around a 11 460 mm diam. circle will give an angle of 26° (26 × 100)

70 mm measured around a 229.2 mm diam. circle will give an angle of 35° (35 × 2)

228 mm measured around a 343.8 mm diam. circle will give an angle of 76° (76 × 3)

276 mm measured around a 687.6 mm diam. circle will give an angle of 46° (46 × 6)

108 mm measured around a 1031.4 mm diam. circle will give an angle of 12° (12 × 9)

By using this method, the scope is unlimited and, when combined with geometric construction, it is very accurate.

Note For a very accurate method of constructing fine angles, see Chapter 6.

Problem

To transfer any given angle to a given point on
another line.

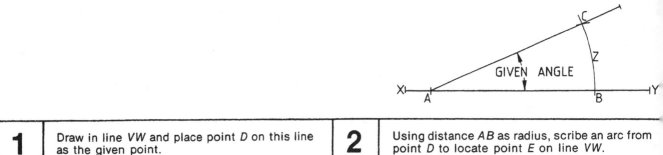

1	Draw in line *VW* and place point *D* on this line as the given point.

2	Using distance *AB* as radius, scribe an arc from point *D* to locate point *E* on line *VW*.

3	Using chord distance *BC* as radius and point *E* as centre, scribe an arc across first arc to locate point *F*.

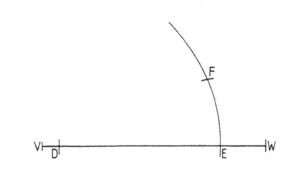

4	Join points *D* and *F* and the angle *FDE* will be equal to the angle *CAB*.

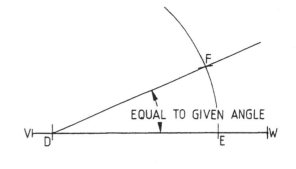

Practical examples

The following are from workshop drawings of angle construction. These are required when marking out a gusset on a structural steel wall truss.

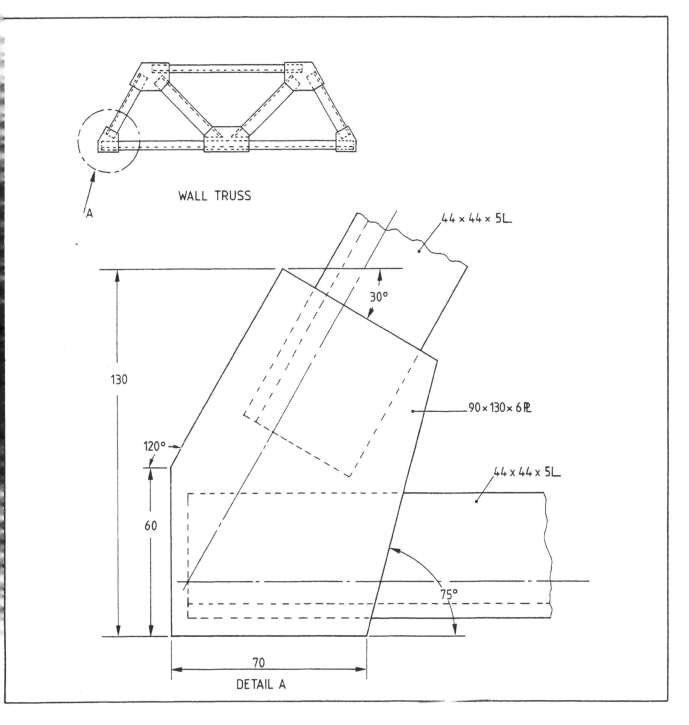

WALL TRUSS

A

44 × 44 × 5L

30°

90 × 130 × 6 ℞

130

120°

44 × 44 × 5L

60

75°

70

DETAIL A

The following examples show construction of
angles taken from workshop drawings.

15° angle on chain guard side plate

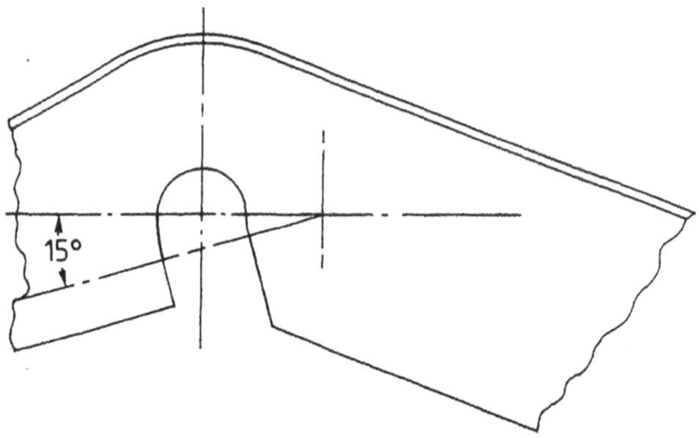

30° angle on hopper gate

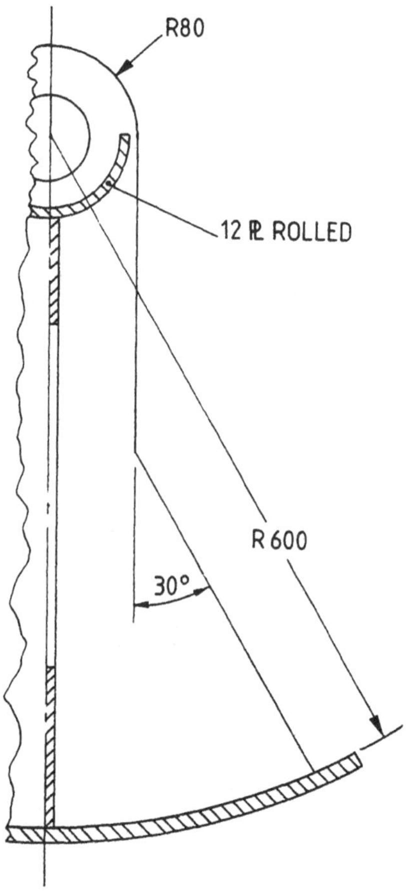

R80

12 ℞ ROLLED

15°

R 600

30°

40° angle on chain guard jockey plate

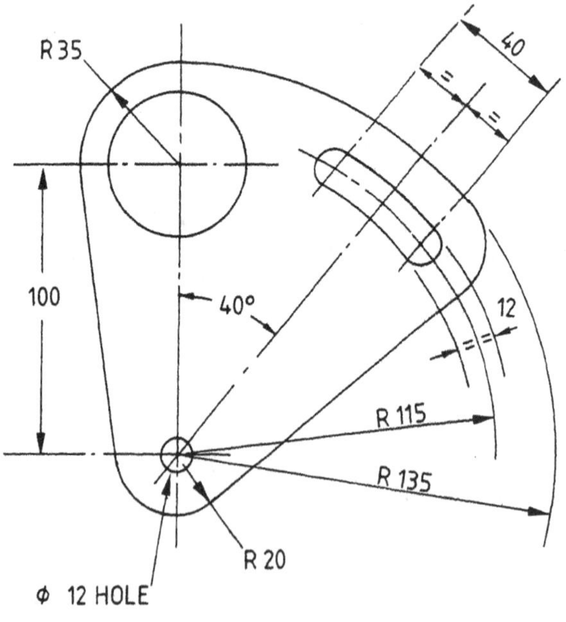

R 35

40

100

40°

12

R 115

R 135

R 20

Φ 12 HOLE

Exercises for constructing and transferring angles

1. Construct a 90° angle off a given line *AB* from the given point *C*.

A ⊢―――――――――――――――⊣B
   C

2. Construct a 45° angle off a given line *DE*.

D ⊢―――――――――――――――⊣E

3. Construct a 60° angle off a given line *FG*.

F ⊢―――――――――――――――⊣G

4. Construct a 30° angle off a given line *HJ*.

H ⊢―――――――――――――――⊣J

Exercises (cont'd)

5. Construct a 20° angle off a given line *AB*.

A ├────────────────────┤ B

6. Construct a 75° angle off a given line *CD*.

C ├────────────────────┤ D

7. Construct a 15° angle off a given line *EF*.

E ├────────────────────┤ F

8. Construct a 126° angle on a given 62 mm diam. semicircle.

Exercises (cont'd)

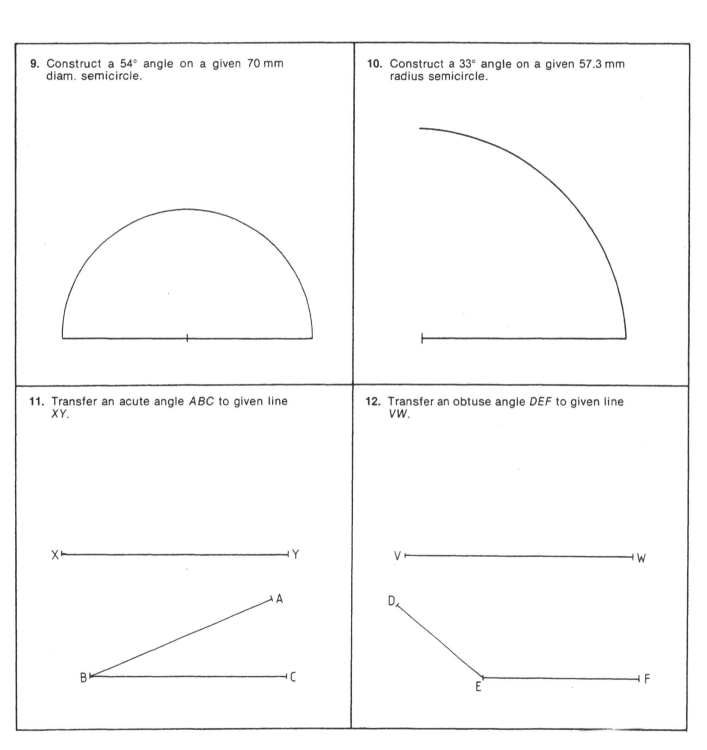

9. Construct a 54° angle on a given 70 mm diam. semicircle.

10. Construct a 33° angle on a given 57.3 mm radius semicircle.

11. Transfer an acute angle *ABC* to given line *XY*.

12. Transfer an obtuse angle *DEF* to given line *VW*.

NOTES

5

Constructing tangents and locating tangent points

The boilermaker marker-off needs to be able to correctly mark out tangents and be able to correctly plot the points of tangency.

The correct plotting of the points of tangency is important for calculation of lengths of material, positioning of the section to be formed and the smooth appearance of the finished job.

Tangents are commonly found on many jobs where straight surfaces meet and the corner is rounded, where straight surfaces merge into curved surfaces, and where two curved surfaces merge together.

This chapter should guide you to a better understanding of tangents and how to correctly plot the points of tangency. All the exercises are simplified, being shown in a step-by-step format.

Note

Do not measure from the diagrams except in exercises, because many figures are indication diagrams only and are not drawn to scale. All measurements on the diagrams are in millimetres unless stated otherwise.

Problem

To scribe an arc tangent to two straight lines at 90° to each other and to locate the points of tangency.

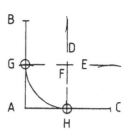

1 (a) Let the two lines, *BA* and *AC*, be at 90°. (b) Scribe arcs using the given distance as radius and points *B*, *A* and *C* as centres.	**2** Draw in lines *D* and *E* tangent to the arcs and intersecting at point *F*.

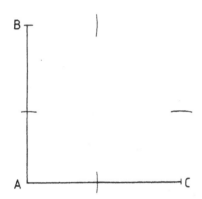

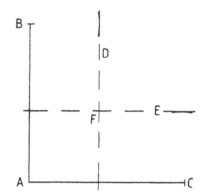

3 Using point *F* as centre and given distance as radius, scribe an arc tangent to lines *BA* and *AC*.	**4** The points of tangency will be points *G* and *H* and they are located by drawing a line from centre point *F* at 90° to lines *BA* and *AC*.

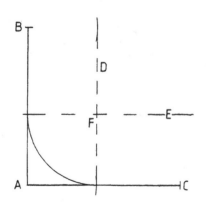

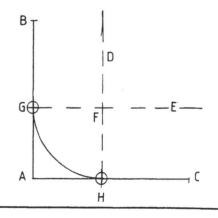

Problem

To scribe an arc tangent to two straight lines forming an acute angle and to locate the points of tangency.

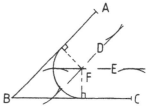

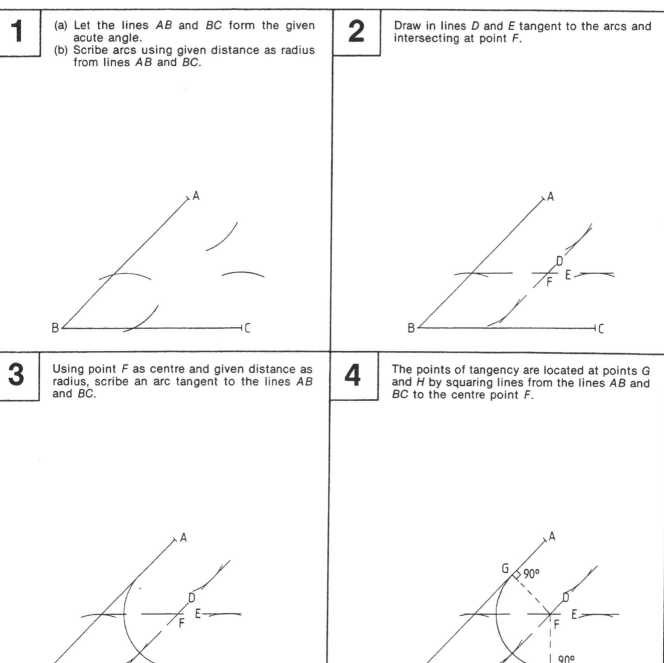

1

(a) Let the lines *AB* and *BC* form the given acute angle.

(b) Scribe arcs using given distance as radius from lines *AB* and *BC*.

2

Draw in lines *D* and *E* tangent to the arcs and intersecting at point *F*.

3

Using point *F* as centre and given distance as radius, scribe an arc tangent to the lines *AB* and *BC*.

4

The points of tangency are located at points *G* and *H* by squaring lines from the lines *AB* and *BC* to the centre point *F*.

Problem

To scribe an arc tangent to two straight lines forming an obtuse angle and to locate the points of tangency.

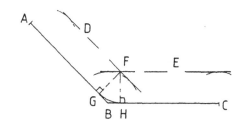

1	(a) Let the lines *AB* and *BC* form the given obtuse angle. (b) Scribe arcs using given distance as radius from lines *AB* and *BC*.

2	Draw in lines *D* and *E* tangent to the arcs and intersecting at point *F*.

3	Using point *F* as centre and given distance as radius, scribe an arc tangent to the lines *AB* and *BC*.

4	The points of tangency are located at points *G* and *H* by squaring lines from the lines *AB* and *BC* to the centre point *F*.

Problem

To scribe an arc tangent to a straight line and a given circle.

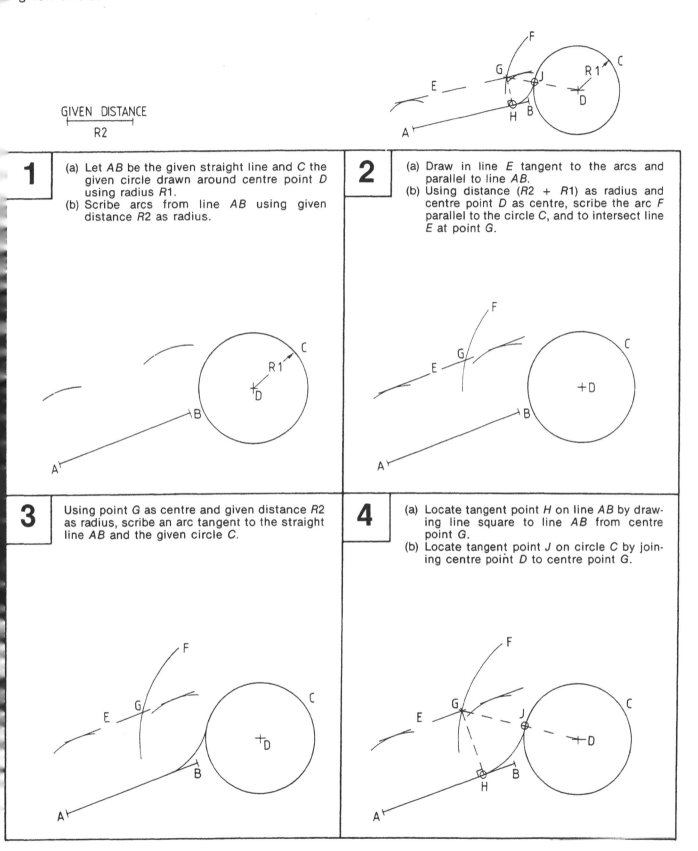

GIVEN DISTANCE
R2

1
(a) Let *AB* be the given straight line and *C* the given circle drawn around centre point *D* using radius *R*1.
(b) Scribe arcs from line *AB* using given distance *R*2 as radius.

2
(a) Draw in line *E* tangent to the arcs and parallel to line *AB*.
(b) Using distance (*R*2 + *R*1) as radius and centre point *D* as centre, scribe the arc *F* parallel to the circle *C*, and to intersect line *E* at point *G*.

3
Using point *G* as centre and given distance *R*2 as radius, scribe an arc tangent to the straight line *AB* and the given circle *C*.

4
(a) Locate tangent point *H* on line *AB* by drawing line square to line *AB* from centre point *G*.
(b) Locate tangent point *J* on circle *C* by joining centre point *D* to centre point *G*.

Problem

To scribe an external arc of a given radius tangent to two given circles and to locate the points of tangency.

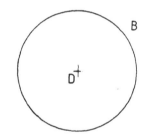

1

(a) Draw the two given circles A and B from centre points C and D using radii R1 and R2.

(b) Using point C as centre and the given distance R3 + R1 as radius, scribe the arc E parallel to circle A.

2

Using point D as centre and the given distance R3 + R2 as radius, scribe the arc F parallel to circle B and intersecting arc E at point G.

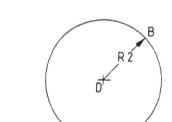

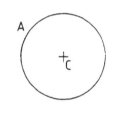

3

Using point G as centre and the given distance R3 as radius, scribe an arc tangent to both the circles A and B.

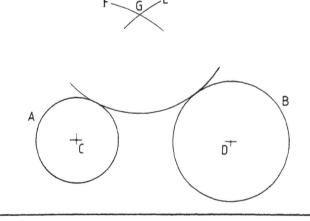

4

(a) Locate tangent point H on circle A by joining centre points C and G.

(b) Locate tangent point J on circle B by joining centre points D and G.

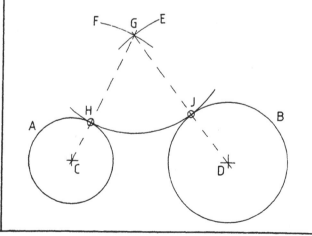

74

Problem

To scribe an internal arc of a given radius tangent to two given circles and to locate the points of tangency.

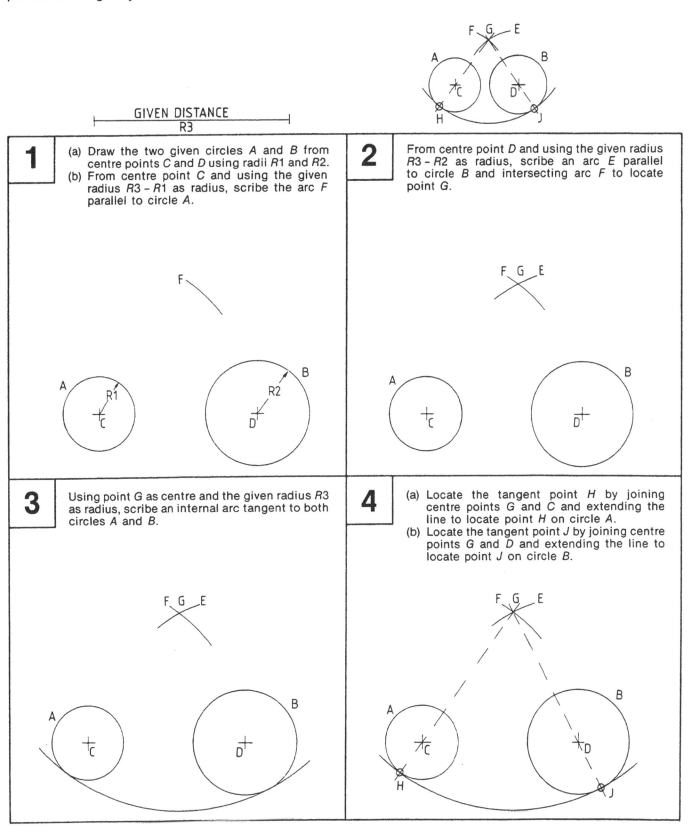

GIVEN DISTANCE
R3

1
(a) Draw the two given circles *A* and *B* from centre points *C* and *D* using radii *R1* and *R2*.
(b) From centre point *C* and using the given radius *R3 – R1* as radius, scribe the arc *F* parallel to circle *A*.

2
From centre point *D* and using the given radius *R3 – R2* as radius, scribe an arc *E* parallel to circle *B* and intersecting arc *F* to locate point *G*.

3
Using point *G* as centre and the given radius *R3* as radius, scribe an internal arc tangent to both circles *A* and *B*.

4
(a) Locate the tangent point *H* by joining centre points *G* and *C* and extending the line to locate point *H* on circle *A*.
(b) Locate the tangent point *J* by joining centre points *G* and *D* and extending the line to locate point *J* on circle *B*.

Problem

To scribe an internal-external arc at a given radius tangent to two given circles and to locate the points of tangency.

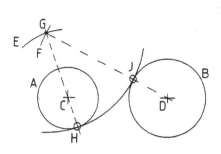

1

(a) Draw the two given circles A and B from centre points C and D using radii R1 and R2.
(b) From centre point C and using radius R3 – R1, scribe the arc E parallel to circle A.

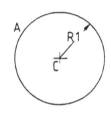
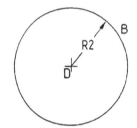

2

From centre point D and using radius R2 + R3, scribe the arc F parallel to circle B and intersecting arc E to locate point G.

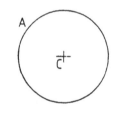
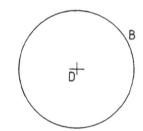

3

Using point G as centre and the given distance R3 as radius, scribe the external-internal arc tangent to both circles A and B.

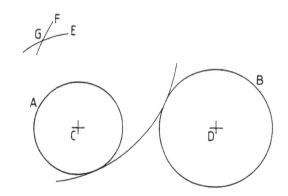

4

(a) Locate the tangent point H by joining centre points G and C and extending the line to cut circle A at point H.
(b) Locate the tangent point J by joining centre points G and D to cut circle B at point J.

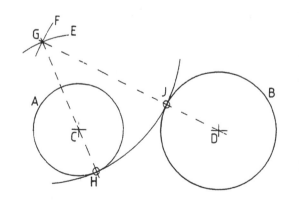

Practical examples

This workshop drawing of the elevation of a chain guard shows the use and need for tangent points. They are used for marking out the shape of the side plate, marking correct shape of slotted holes, calculating the length of the wrapper plate and marking the positions and lengths of rolled sections of the wrapper plate.

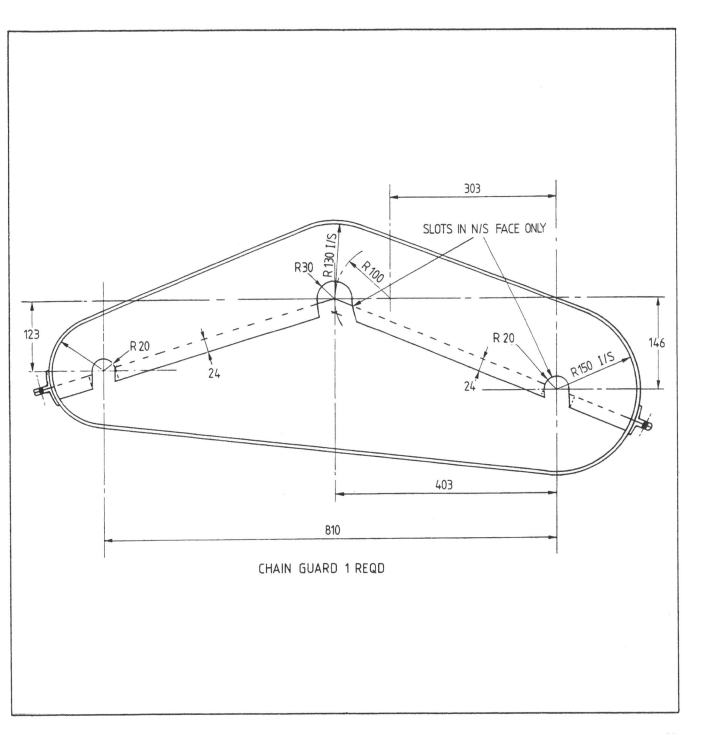

CHAIN GUARD 1 REQD

This girder clamp plate is a good example to
show the use of tangent points.

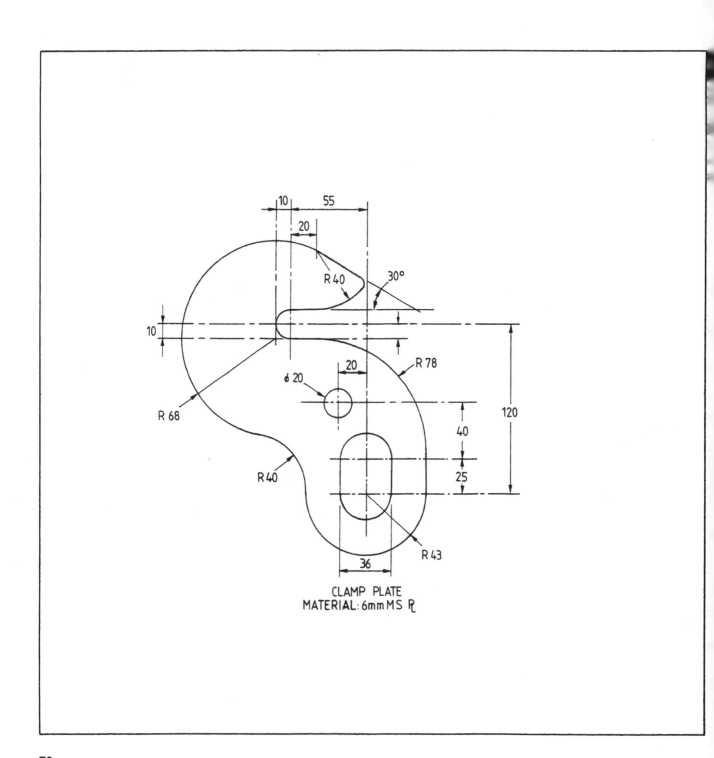

CLAMP PLATE
MATERIAL: 6mm MS Pl

This curved cam plate shows tangent points
used where curved lines are merging with other
curved lines.

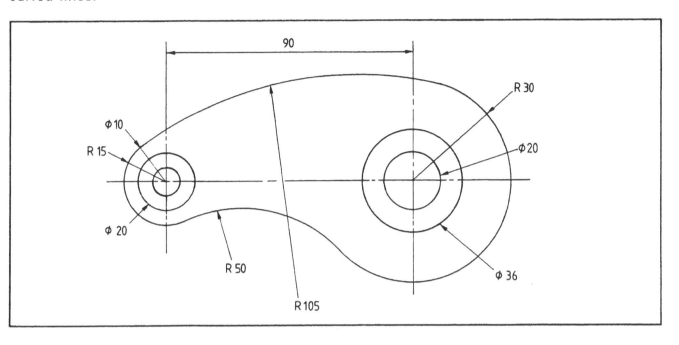

This swinging arm link bar shows the use of
tangent points where straight lines merge into
curved lines.

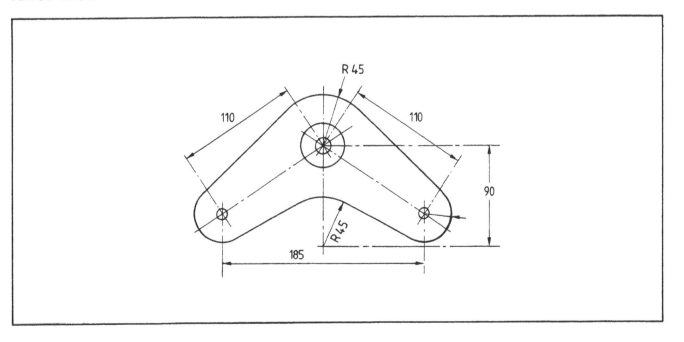

Exercises for drawing arcs to locate tangent points

1. Draw an arc of 40 mm radius, tangent to two straight lines at right angles. Locate the points of tangency.

2. Draw an arc of 20 mm radius tangent to two straight lines forming an acute angle. Locate the points of tangency.

3. Draw an arc of 30 mm radius tangent to two straight lines forming an obtuse angle. Locate the points of tangency.

4. Draw an arc of 20 mm radius tangent to a circle and a straight line. Locate the points of tangency.

Exercises (cont'd)

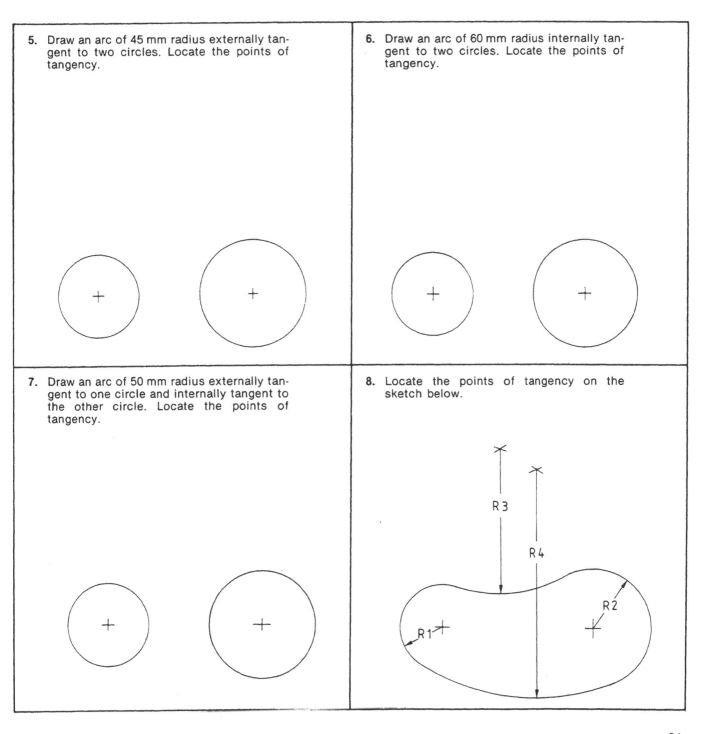

5. Draw an arc of 45 mm radius externally tangent to two circles. Locate the points of tangency.

6. Draw an arc of 60 mm radius internally tangent to two circles. Locate the points of tangency.

7. Draw an arc of 50 mm radius externally tangent to one circle and internally tangent to the other circle. Locate the points of tangency.

8. Locate the points of tangency on the sketch below.

R 3

R 4

R 2

R 1

NOTES

6

Calculations required for marking off by boilermakers

In this chapter you are supplied with enough information to assist you in your calculations for marking off on layouts and jobs.

In these modern times electronic calculators have come into their own to assist the tradesman with calculations for layouts and in some instances the ease of calculations with calculators has even eliminated the need to lay out some types of jobs.

This chapter is set out in various sections with each section having formulae, reference tables, examples and exercises for you to work out with reference to our examples.

The various sections are:

Note

Do not measure from the diagrams except in exercises, because many figures are indication diagrams only and are not drawn to scale. All measurements on the diagrams are in millimetres unless stated otherwise.

6.1 Definitions of mathematic and geometric signs and symbols

Sign or symbol	Definition	Example of use
+	*Plus* The sign of addition.	$20 + 5 = 25$
−	*Minus* The sign of subtraction.	$12 - 4 = 8$
×	*Multiply* The sign of multiplication.	$4 \times 4 = 16$
÷	*Divide* The sign of division.	$9 \div 3 = 3$
=	*Equal sign* The sign of equality; equal to.	$6 + 6 = 12$
≠ or ≠	*Not equal to.*	$16 \neq 21$
>	*Greater than* The value written before the sign is greater than the value written after the sign.	$4 > 3$
≯	*Not greater than* The value written before the sign is not greater than the value written after the sign.	$\$1.20 \ngtr 120c$
<	*Less than* The value written before the sign is less than the value written after the sign.	$9 < 10$
≮	*Not less than* The value written before the sign is not less than the value written after the sign.	$37c \nless \$0.37$
$\sqrt{}$	*Square root sign* The answer when multiplied by itself will equal the original number.	$\sqrt{25} = 5$ Thus 5 is the square root of 25.
6^2	The small superscript ² above any number means that this number must be multiplied by itself (squared).	$6 \times 6 = 36$
6^3	The small superscript ³ above any number means that this number is to be cubed, or multiplied by itself and this product multiplied by the original number.	$6 \times 6 \times 6 = 216$ or $(6 \times 6 = 36) \times 6 = 216$
: :: :	: is to (ratio) :: as (proportion) 4 is to 8 as 5 is to 10.	$4 : 8 :: 5 : 10$
π	*Pi* A transcendental number, approximately 3.14159, representing the ratio of the circumference to the diameter of a circle and appearing as a constant in a wide range of mathematical problems. In boilermaking we use $\pi = 3.1416$.	$\pi \times$ diameter = the circumference of a circle $\dfrac{\pi D^2}{4}$ = area of a circle

ID — Diameter measured on inside of cylinder. — ID $\times \pi$ = inside circumference

MD — Diameter measured to centre of plate thickness of cylinder (neutral axis). — MD $\times \pi$ = mean circumference

OD — Diameter measured on outside of cylinder. — OD $\times \pi$ = outside circumference

Sign or symbol	Definition	Example of use

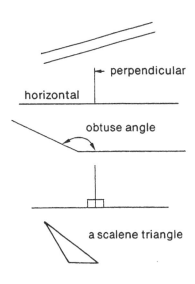

⊩ *Parallel* Lines at equal distance apart or two lines that can never meet.

⊥ *Perpendicular* A line at right angles to the horizontal.

∠ *Angle* The figure formed by two lines diverging from a common point.

⌐ *Right angle* Formed by two lines meeting at 90°.

△ *Triangle* A three-sided flat figure.

( ) *Parentheses* Used to enclose a sum, product, or other expression considered or treated as a collecting entity in a mathematical problem.

length $= 3.1416 \times$
$(D + T + \frac{F}{3})$

[ ] *Brackets* Symbols used to enclose a mathematical problem.

mass $= \pi\,[3(3.3 + 8.1) - 2.7]$

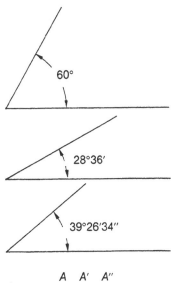

° *Degree* A unit of angular measure equal in magnitude to the central angle subtended by ⅟₃₆₀ of the circumference of a circle.

′ *Minute* A unit of angular measurement equal to ⅟₆₀ of a degree.

″ *Second* A unit of angular measurement equal to ⅟₆₀ of a minute.

′ and ″ These are also used in transformation of labelled points.

$A \quad A' \quad A''$

% *Per cent* Per hundred; for or out of each hundred, e.g. one-quarter is 25%.

$0.98 = 98\%$

@ *At* Used in material or cutting lists for abbreviation.

4 @ 200 mm long

6.2 Formulae for calculations

Perimeter and area		Perimeter = P	Area = A
Square		P = 4S	A = S²
Rectangle		P = 2(L + B)	A = L × B
Triangle (perpendicular height given)		P = Sum of 3 sides	$A = \dfrac{B \times PH}{2}$
Scalene triangle		P = a + b + c	$A = \sqrt{S(S - a)(S - b)(S - c)}$ $S = \dfrac{a + b + c}{2}$
Equilateral triangle		P = 3S	A = 0.433 S²
Circle		C = π D D = 0.3183 C	$A = \dfrac{\pi D^2}{4}$ For mean circum. π = 3.1416

Pythagoras' Theorem

$c = \sqrt{a^2 + b^2}$

$a = \sqrt{c^2 - b^2}$

$b = \sqrt{c^2 - a^2}$

Square plate from circular plate

Area of largest square plate that can be cut from a circular plate $A = \dfrac{D^2}{2}$

Length of side of largest square plate
S = 0.7071 D
D = 1.4142 S

Surface area and volume	Surface area = SA Length of welding = WL (tank fully enclosed)	Volume = V
Rectangular or square tank	$SA = 2(L \times B) + 2(L \times H) + 2(B \times H)$ $WL = 4L + 4B + 4H$	$V = L \times B \times H$
Cylindrical tank	$SA = \pi\,DH + \dfrac{2\pi D^2}{4}$ $WL = 2\pi D + H$	$V = \dfrac{\pi D^2 \times H}{4}$
Right cone	$SA = \dfrac{\pi D \times SH}{2} + \dfrac{\pi D^2}{4}$ $WL = \pi D + SH$ $SH = \sqrt{H^2 + R^2}$	$V = \dfrac{\pi D^2 \times H}{4 \times 3}$
Sphere	$SA = \pi D^2$	$V = \dfrac{\pi D^3}{6}$
Annular plate	$SA = \pi \times MD \times W$ or $SA = \dfrac{\pi}{4}(D^2 - d^2)$	

Cubic contents

1 m³ contains 1000 litres.
1 litre of water has a mass of 1 kg.
∴ 1 m³ of water has a mass of 1000 kg.

Note $\pi = 3.1416$

6.3 Diameter and circumference of cylinders

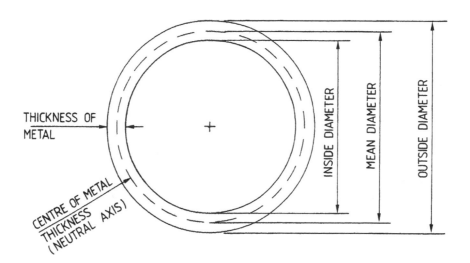

When forming metal into a curved surface, boilermakers must always be aware of the need to allow for the centre of the metal remaining constant (neutral axis), the outside surface stretching and the inside surface compressing.

Calculating the correct length of metal plate prior to rolling is based on the use of the mean diameter.

The above diagram shows:
1. Mean diameter = inside diameter + 1 plate thickness
   written as MD = ID + T.

2. Mean diameter = outside diameter − 1 plate thickness
   written as MD = OD − T.

ID + 2T = OD   MD + T = OD
OD − 2T = ID   MD − T = ID
OD − T = MD   (2 × IR) + T = MD
ID + T = MD   (2 × OR) − T = MD

Remember you must always use *mean diameter* when calculating the *mean circumference* of any curved jobs made from metal.

Example

Calculate the length of mild steel flat plate required to form the cylinder drawn below.

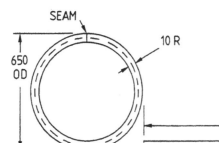

Length = MD × π = MC before rolling

MD = OD − T
   = 650 − 10
   = 640 mm
MC = MD × π
   = 640 × 3.1416
   = 2010.624 mm
     say, 2011 mm or 2.011 m

Thus a piece of 10 mm thick mild steel plate, 2011 mm long, would be required to roll up a cylinder measuring 630 mm ID or 650 mm OD.

LENGTH = MD × π = MC BEFORE ROLLING

Exercise

Calculate the mean circumference of a 360 mm OD cylinder made from 10 mm mild steel plate.

$$MD = OD - T \qquad MC = MD \times \pi$$
$$= \qquad\qquad =$$
$$= \quad mm \qquad\qquad = \quad mm$$
$$\qquad\qquad\qquad say, \quad mm$$
$$\qquad\qquad\qquad or \quad m$$

Example

Calculate the length of mild steel plate required to form a 560 mm ID cylinder with a plate thickness of 8 mm.

$$MD = ID + T$$
$$= 560 + 8$$
$$= 568 \ mm$$
$$MC = MD \times \pi$$
$$= 568 \times 3.1416$$
$$= 1784.4288 \ mm$$
$$say, 1784 \ mm \ or \ 1.784 \ m$$

Exercise

Calculate the MC of a 330 mm ID cylinder made from 6 mm mild steel plate.

$$MD = ID + T$$
$$=$$
$$= \quad mm$$
$$MC = MD \times \pi$$
$$=$$
$$= \quad mm$$
$$say, \quad mm \ or \quad m$$

Example

Calculate the length of mild steel plate required to form a cylinder having a 220 mm inside radius and a thickness of 6 mm.

$$MD = (2 \times IR) + T$$
$$= (2 \times 220) + 6$$
$$= 440 + 6$$
$$= 446 \ mm$$
$$MC = MD \times \pi$$
$$= 446 \times 3.1416$$
$$= 1401.1536 \ mm$$
$$say, 1401 \ mm \ or \ 1.401 \ m$$

Exercise

Calculate the MC of a cylinder having a 280 mm inside radius and a thickness of 8 mm.

$$MD = (2 \times IR) + T$$
$$=$$
$$=$$
$$= \quad mm$$
$$MC = MD \times \pi$$
$$=$$
$$= \quad mm$$
$$say, \quad mm \ or \quad m$$

To convert a circumference back to a diameter, reverse the calculation by dividing the circumference by π.

Example

Calculate the outside diameter of a 6 mm thick mild steel cylinder having an outside circumference of 1724 mm.

$$OD = OC \div \pi$$
$$= 1724 \div 3.1416$$
$$= 548.764\,96 \ mm$$
$$say, 549 \ mm$$

Exercise

Calculate the outside diameter of an 8 mm thick mild steel cylinder having an outside circumference of 1627 mm.

$$OD = OC \div \pi$$
$$=$$
$$=$$
$$say, \quad mm$$

Example

Calculate the inside diameter of a 6 mm thick mild steel cylinder having an outside circumference of 1226 mm.

$$ID = (OC \div \pi) - 2T$$
$$= (1226 \div 3.1416) - 2 \times 6$$
$$= 390.247 - 12$$
$$= 378.247 \ mm$$
$$say, 378 \ mm$$

Exercise

Calculate the inside diameter of a 5 mm thick mild steel cylinder having an outside circumference of 920 mm.

$$ID = (OC \div \pi) - 2T$$
$$=$$
$$=$$
$$=$$
$$say, \quad mm$$

Table 6.1 Conversion of diameters to circumferences (all measurements in millimetres to 3 decimal places)

(a) Diameters 1 mm to 9 mm progressing by millimetres

Diam.	1 mm	2 mm	3 mm	4 mm	5 mm	6 mm	7 mm	8 mm	9 mm
Circ.	3.1416	6.2832	9.4248	12.5664	15.708	18.8496	21.9912	25.1328	28.2744

(b) Diameters 10 mm to 90 mm progressing by tens

Diam.	10 mm	20 mm	30 mm	40 mm	50 mm	60 mm	70 mm	80 mm	90 mm
Circ.	31.416	62.832	94.248	125.664	157.08	188.496	219.912	251.328	282.744

(c) Diameters 100 mm to 900 mm progressing by hundreds

Diam.	100 mm	200 mm	300 mm	400 mm	500 mm	600 mm	700 mm	800 mm	900 mm
Circ.	314.16	628.32	942.48	1256.64	1570.8	1884.96	2199.12	2513.28	2827.44

(d) Diameters 1000 mm to 10 000 mm progressing by thousands

Diam.	1000 mm	2000 mm	3000 mm	4000 mm	5000 mm	6000 mm	7000 mm	8000 mm	9000 mm	10 000 mm
Circ.	3141.6	6283.2	9424.8	12 566.4	15 708	18 849.6	21 991.2	25 132.8	28 274.4	31 416

Table 6.1 is a quick reference for converting diameters to circumferences, from 1 mm as the smallest diameter progressing by millimetres up to 10 999 mm diameter.

The following examples show you how to use the table for quick accurate results.

Example

Convert 542 mm diameter to a circumference.

```
500 mm from Table 6.1(c) = 1570.8 mm
 40 mm from Table 6.1(b) =  125.664 mm
  2 mm from Table 6.1(a) =    6.2832 mm
542 mm diameter          = 1702.7472 mm
                             circumference
```

Exercise

Convert 379 mm diameter to a circumference.

```
300 mm from Table 6.1(c) =
 70 mm from Table 6.1(b) =
  9 mm from Table 6.1(a) =
379 mm diameter          =          mm
                              circumference
```

Example

Convert 1897 mm diameter to a circumference.

```
1000 mm from Table 6.1(d) = 3141.6 mm
 800 mm from Table 6.1(c) = 2513.28 mm
  90 mm from Table 6.1(b) =  282.744 mm
   7 mm from Table 6.1(a) =   21.9912 mm
1897 mm diameter          = 5959.6152 mm
                              circumference
```

Exercise

Convert 3283 mm diameter to a circumference.

```
3000 mm                   =
 200 mm                   =
  80 mm                   =
   3 mm                   =
3283 mm diameter          =          mm
                              circumference
```

Table 6.2 Conversion of circumferences to diameters (all measurements in millimetres to 3 decimal places)

(a) Circumferences 1 mm to 9 mm progressing by millimetres

Circ.	1 mm	2 mm	3 mm	4 mm	5 mm	6 mm	7 mm	8 mm	9 mm
Diam.	0.318	0.636	0.955	1.273	1.592	1.910	2.228	2.546	2.865

(b) Circumferences 10 mm to 90 mm progressing by tens

Circ.	10 mm	20 mm	30 mm	40 mm	50 mm	60 mm	70 mm	80 mm	90 mm
Diam.	3.183	6.366	9.549	12.732	15.915	19.099	22.282	25.465	28.648

(c) Circumferences 100 mm to 900 mm progressing by hundreds

Circ.	100 mm	200 mm	300 mm	400 mm	500 mm	600 mm	700 mm	800 mm	900 mm
Diam.	31.831	63.662	95.493	127.324	159.154	190.985	222.816	254.647	286.478

(d) Circumferences 1000 mm to 10 000 mm progressing by thousands

Circ.	1000 mm	2000 mm	3000 mm	4000 mm	5000 mm	6000 mm	7000 mm	8000 mm	9000 mm	10 000 mm
Diam.	318.309	636.618	954.927	1273.236	1591.546	1909.855	2228.164	2546.473	2864.782	3183.091

Table 6.2 is a quick reference for converting circumferences to diameters, from 1 mm as the smallest circumference and progressing by millimetres up to 10 999 mm circumference.

The following examples show how to use the table for quick accurate results.

Example

Convert 1703 mm circumference to a diameter.

1000 mm from Table 6.2(d) =	318.309 mm	
700 mm from Table 6.2(c) =	222.816 mm	
3 mm from Table 6.2(a) =	.955 mm	
1703 mm circumference	= 542.080 mm diameter say, 542 mm	

Exercise

Convert 879 mm circumference to a diameter.

800 mm from Table 6.2(c) =	
70 mm from Table 6.2(b) =	
9 mm from Table 6.2(a) =	
879 mm circumference =	mm diameter say, mm

Example

Convert 5960 mm circumference to a diameter.

5000 mm from Table 6.2(d) =	1591.546 mm
900 mm from Table 6.2(c) =	186.478 mm
60 mm from Table 6.2(b) =	19.099 mm
5960 mm circumference =	1797.123 mm diameter say, 1797 mm

Exercise

Convert 5637 mm circumference to a diameter.

5000 mm	=
600 mm	=
30 mm	=
7 mm	=
5637 mm circumference =	mm diameter say, mm

6.4 Circumference of an ellipse

There is no absolutely correct method to calculate the circumference of an ellipse.

The usually accepted method for calculating the circumference of an ellipse is to add the minor and major axes, divide the result by 2, then multiply by π (3.1416).

$$C = \left(\frac{\text{minor axis} + \text{major axis}}{2} \right) \times 3.1416$$

The above formula has been said to be close enough to correct for the generally used elliptical shapes, but it is a long way out when calculating the long thin ellipse.

For all types of elliptical shapes, we recommend a formula (taken from I. J. & H. Haddon's book *A Practical Treatise for Boilermakers*) which comes out with an answer that is much closer to correct for all types of ellipses. That formula is as follows:

To four times the hypotenuse of a right angle contained in a quarter of an ellipse, add 0.6264 of half the minor axis.

$$C = (4 \times \text{length of line } A) + (0.6264 \times \tfrac{1}{2} \text{ minor axis})$$

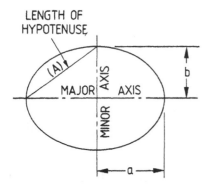

LENGTH OF HYPOTENUSE

There is a third formula which has been used for many years.

$$C = \pi \left[\frac{3(a + b)}{2} - \sqrt{ab} \right]$$

The difference between these two method and the actual physical measurement of a ellipse is shown below.

Calculate the circumference of an ellipse having a major axis of 610 mm and a minor axis of 305 mm.

(a) Usually accepted method

$$C = \left(\frac{\text{major axis} + \text{minor axis}}{2} \right) \times \pi$$

$$= \left(\frac{610 + 305}{2} \right) \times 3.1416$$

$$= \frac{915}{2} \times 3.1416$$

$$= 457.5 \times 3.1416$$

$$= 1437.282 \text{ mm}$$
$$\text{say, } 1437 \text{ mm}$$

(b) Our recommended method

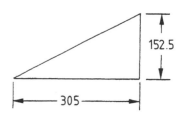

$$\text{Length of hypotenuse} = \sqrt{305^2 + 152.5^2}$$

$$= \sqrt{93\,025 + 23\,256.25}$$

$$= \sqrt{116\,281.25}$$

$$= 341.000\,36 \text{ mm}$$

$$C = (4 \times \text{hypotenuse}) + (0.6264 \times 0.5 \text{ of minor axis})$$

$$= (4 \times 341) + (0.6264 \times 152.5)$$

$$= 1364 + 95.526$$

$$= 1459.526 \text{ mm}$$
$$\text{say, } 1460 \text{ mm}$$

The layout of this ellipse when accurately measured is 1470 mm, making our recommended method much closer than the usually accepted method.

The usually accepted calculation is 33 mm short and our recommended method is 10 mm short of the measured job. The third formula gives an answer which is 9 mm longer than the measured job.

As an extreme example, let it be required to find the circumference of an ellipse with the major axis 635 mm and the minor axis 127 mm.

(a) Usually accepted method

$$C = \left(\frac{\text{major axis } + \text{ minor axis}}{2} \right) \times \pi$$

$$= \left(\frac{635 + 127}{2} \right) \times 3.1416$$

$$= \frac{762}{2} \times 3.1416$$

$$= 381 \times 3.1416$$

$$= 1196.9496 \text{ mm}$$
$$\text{say, } 1197 \text{ mm}$$

(b) Our recommended method

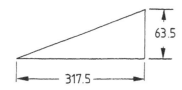

Length of hypotenuse $= \sqrt{317.5^2 + 63.5^2}$

$$= \sqrt{100\,806.25 + 4032.25}$$

$$= \sqrt{104\,838.5}$$

$$= 323.787\,73 \text{ mm}$$

$$C = (4 \times \text{hypotenuse}) + (0.6264 \times 63.5)$$
$$= (4 \times 323.787\,73) + 39.7764$$
$$= 1295.1509 + 39.7764$$
$$= 1334.9273 \text{ mm}$$
$$\text{say, } 1335 \text{ mm}$$

As you can see, there is 138 mm difference between the two methods. The accurate measurement of the ellipse is 1330.325 mm, say 1330 mm, showing that our recommended method is within 5 mm of the correct measurement. The usually accepted method shows up badly here because the answer is less than double the length of the major axis (635 + 635 = 1270 mm).

This proves that the flatter the elliptical shape, the further out the usually accepted method becomes. The third formula gives an answer of 1349 mm, which is 19 mm longer than the measured length.

An examination of the following ellipse shows clearly that the circumference of an ellipse must be greater than 4 times the length of the hypotenuse line *A* and less in length than the perimeter of the rectangle *BCDE*.

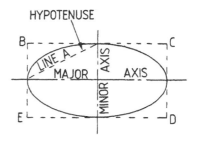

These comparisons of the three methods highlight the discrepancy between the usually accepted method (used because of simplicity) and the method we recommend. Where the accurate measurement of a layout of an ellipse or the accurate measurement around an elliptical bulkhead or division plate is not practical and the circumference has to be calculated, then the facts and figures we have presented must lead you to using our recommended method to enable a closer calculated circumference on all generally flatter ellipses under 60°. The usually accepted method would be best suited to rounder shaped ellipses above 60°.

(a) Usually recommended method

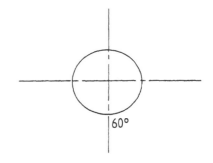

(b) Our recommended method

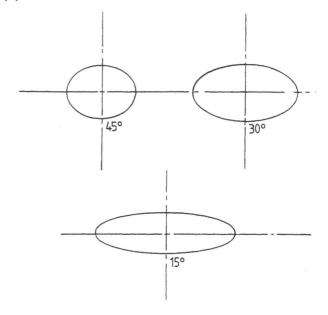

6.5 Circumference of cylinders for telescopic fit-up (lap joints)

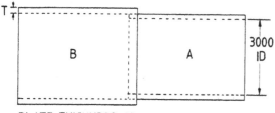

PLATE THICKNESS: 10mm

When calculating the length of plate for cylinders or rings to fit over or inside existing cylinders or rings with a lap joint (as drawn above), an allowance must be made so that they are a working fit. For example, a 30 mm bolt will not fit a 30 mm hole, so the hole is drilled to, say, 32 mm to allow the bolt to fit into the hole. The difference between the MD of the cylinder B and the MD of the cylinder A is twice the thickness of plate (T), so the difference in the circumferences of the two cylinders is 3.1416×2 T (no allowance yet for working fit).

To allow for a working fit, the circumference of cylinder B is calculated by using the formula, MC of cylinder $A + (3.25 \times 2$ T), which is MC of cylinder $A + (6.5 \times$ T).

$6.5 \times T$ is the allowance for *close fitting work*.

$7.5 \times T$ is the allowance for *general work*.

For a cylinder or ring to fit *over* an existing cylinder, add the allowance.

For a cylinder or ring to fit *into* an existing cylinder, subtract the allowance.

Example

Calculate the length of plates required for cylinders A and B from the drawing below, using the allowance for close fitting work.

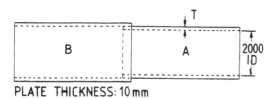

PLATE THICKNESS: 10 mm

MD of cylinder A = ID + T
$= 2000 + 10$
$= 2010$ mm

MC of cylinder A = MD $\times \pi$
$= 2010 \times 3.1416$
$= 6314.616$ mm

MC of cylinder B = MC of cylinder $A + 6.5$ T
$= 6314.616 + (6.5 \times 10)$
$= 6314.616 + 65$
$= 6379.616$ mm
say, 6380 mm

Exercise

Calculate the mean circumference of the rings A and B drawn below. Ring A has an ID of 950 mm and rings A and B are 8 mm thick plate. Use allowance for close fitting work.

PLATE THICKNESS: 8mm

MD of ring A = ID + T
$=$
$=$ mm

MC of ring A = MD $\times \pi$
$=$
$=$ mm

MC of ring B = MC of ring $A + 6.5$ T
$=$
$=$
say, mm

Example

Calculate the length of plates required for cylinders A and B drawn below, using the allowance for *general work*.

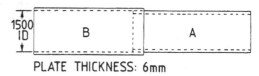

PLATE THICKNESS: 6mm

MD of cylinder B = ID + T
 = 1500 + 6
 = 1506 mm

MC of cylinder B = MD × π
 = 1506 × 3.1416
 = 4731.2496 mm

MC of cylinder A = MC of cylinder B − 7 T
 = 4731.2496 − (7 × 6)
 = 4731.2496 − 42
 = 4689.2496 mm
 say, 4689 mm

Exercise

Calculate the length of plate required for backing ring B to fit into the existing cylinder A as drawn below, using the allowance for *general work*.

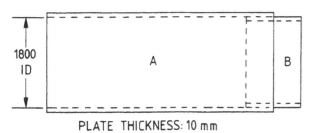

PLATE THICKNESS: 10 mm

MD of cylinder A = ID + T
 =
 =   mm

MC of cylinder A = MD × π
 =
 =   mm

MC of ring B = MC of cylinder A − 7 T
 =
 =
 say,   mm

6.6 Practical method for calculating degrees around circles

These calculations have been explained in Chapter 4. This section consists of examples and exercises. If you have any trouble with these, refer back to Chapter 4.

To calculate a distance for any given degree around any circular surface, the following formula can be used:

$$1° = \frac{d \times \pi}{360}$$

where d = diameter of the surface,
 π = 3.1416,
 360 = no. of degrees in a circle.

Example

Calculate the distance around the outside surface of the tank from centre line D to locate the centre line of the outlet pipe shown on the drawing below.

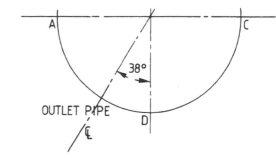

$$38° = \left(\frac{OD \times \pi}{360}\right) \times 38$$

$$= \left(\frac{3300 \times 3.1416}{360}\right) \times 38$$

$$= \left(\frac{10\,367.28}{360}\right) \times 38$$

$$= 28.798 \times 38$$

$$= 1094.324$$

$38°$ = 1094.324 mm on 3300 mm OD tank
 say, 1094 mm

∴ measure 1094 mm from ℄ D around outside surface of the tank to locate ℄ of outlet pipe.

Exercise 1

Calculate the distance around the outside surface of the pipe from centre line A to locate the centre line of the inlet branch as shown on the sketch below.

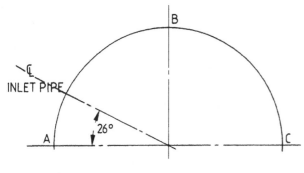

$$26° = \left(\frac{OD \times \pi}{360}\right) \times 26$$

$$=$$

$$=$$

$$=$$

$26° = \qquad$ mm on 920 mm OD pipe

Exercise 2

Calculate the distance around the outside surface of the water tank drawn below from centre line C to locate the centre line of the inlet pipe.

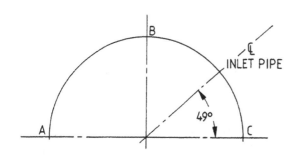

$$49° = \left(\frac{OD \times \pi}{360}\right) \times 49$$

$$=$$

$$=$$

$$=$$

$49° = \qquad$ mm on 18 320 mm OD tank

Check back to Chapter 4 for an alternative method and further reference.

Another method of calculating angles of any magnitude around a circular surface is by using the degree and radians tables with the following formula:

distance around curve = coefficient × radius

Example

Calculate the distance around the 3.048 metre OD cylindrical tank to position centre line of bracket from centre line B as drawn below.

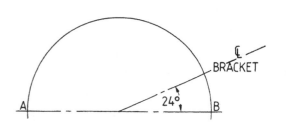

Coefficient of 24° = 0.4189

$$\text{Outside radius} = \frac{OD}{2}$$

$$= \frac{3.048}{2}$$

$$= 1.524 \text{ m}$$

Distance around curve = coefficient × outside radius
= 0.4189 × 1.524
= 0.638 403 6 m
or 638.4 mm

Exercise

Calculate the distance around the 2248 mm OD cylindrical tank to position centre line of outlet pipe from centre line B as drawn below.

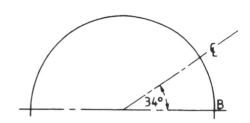

Distance around curve = coefficient × outside radius

$$=$$

$$= \qquad \text{mm}$$

6.7 Calculating hole pitches around a pitch circle diameter (PCD)

Calculation of hole pitches on a PCD are achieved by using the following formula. The answer is the measurement *around the curve*.

$$\text{pitch} = \frac{PCD \times \pi}{\text{no. of holes}}$$

where PCD = pitch circle diameter,
  π = 3.1416,
no. of holes = no. of holes required on PCD.

Note Pitch is measured around circumference *not* across chord.

Example 1

Calculate the pitch of holes of the flange drawn below.

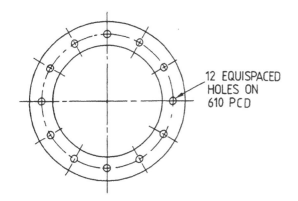

12 EQUISPACED HOLES ON 610 PCD

$$\text{pitch} = \frac{PCD \times \pi}{\text{no. of holes}}$$
$$= \frac{610 \times 3.1416}{12}$$
$$= \frac{1916.376}{12}$$
$$= 159.698 \text{ mm}$$

∴ Measure 159.698 mm *around the curve* to check hole pitch distance. This calculation is very helpful when you need to know the distance for half and quarter pitch to start off either side of a centre line.

Note The number of pitches around a circumference of a circle equals number of holes.

Example 2

31 holes on 1200 mm PCD, calculate pitch.

$$\text{pitch} = \frac{PCD \times \pi}{\text{no. of holes}}$$
$$= \frac{1200 \times 3.1416}{31}$$
$$= 121.61 \text{ mm}$$
say, 122 mm

Exercise 1

Calculate hole pitch on flange with 250 mm PCD and 8 holes.

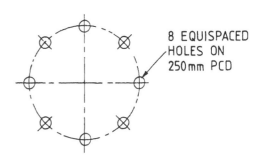

8 EQUISPACED HOLES ON 250 mm PCD

$$\text{pitch} = \frac{PCD \times \pi}{\text{no. of holes}}$$
$$= \underline{\hspace{3cm}}$$
$$=$$
say,   mm

Exercise 2

Calculate the pitch between holes: 34 holes on 1200 mm PCD.

$$\text{pitch} = \frac{PCD \times \pi}{\text{no. of holes}}$$
$$= \underline{\hspace{3cm}}$$
$$=$$
say,   mm

Exercise 3

Calculate the pitch between holes: 176 holes on 4280 mm PCD.

$$\text{pitch} = \frac{\text{PCD} \times \pi}{\text{no. of holes}}$$

$$= \underline{\hspace{3cm}}$$

$$= \underline{\hspace{1.5cm}}$$

say,   mm

Exercise 4

Calculate the pitch between holes: 8 holes on 640 mm PCD.

$$\text{pitch} = \frac{\text{PCD} \times \pi}{\text{no. of holes}}$$

$$= \underline{\hspace{3cm}}$$

$$= \underline{\hspace{1.5cm}}$$

say,   mm

Note Pitch is measured around circumference *not* across chord.

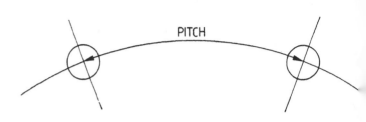

Complete the following table:

Circumference	PCD	No. holes	Pitch
	749	7	
	448	14	
	1568	28	

6.8 Calculating length of material for formed work

When calculating material for formed work, the golden rule to remember is to take all measurements to inside of square sharp bends and use centre line of material (MD) for all curved work.

Example 1

Calculate the length of flat bar required to bend the pipe bracket drawn below.

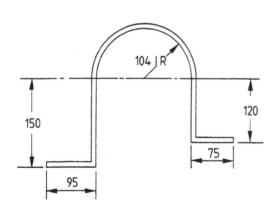

Material: 75 × 10 flat bar

$$\begin{aligned}
\text{Length} &= \text{straight sections} + \text{curved section} \\
&= [(95-10) + (150-10) + (120-10) \\
&\quad + (75-10)] + \frac{\text{MD} \times \pi}{2} \\
&= (85 + 140 + 110 + 65) \\
&\quad + \frac{218 \times 3.1416}{2} \\
&= 400 + \frac{684.8688}{2} \\
&= 400 + 342.4344 \\
&= 742.4344 \text{ mm} \\
&\quad \text{say, 743 mm}
\end{aligned}$$

The piece of flat bar would be marked out as follows ready for forming in press and rolls.

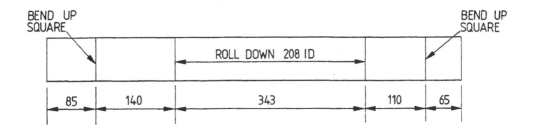

BEND UP SQUARE

BEND UP SQUARE

ROLL DOWN 208 ID

| 85 | 140 | 343 | 110 | 65 |

Example 2

Calculate the length of material required to form the obround manhole stiffener plate drawn below.

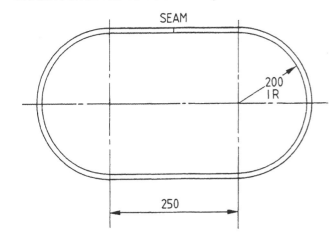

SEAM

200 IR

250

Material: 20 mm plate

Length = straight sections + curved sections
= (250 + 250) + (MD × π)
= 500 + (420 × 3.1416)
= 500 + 1319.472
= 1819.472 mm
say, 1820 mm

The piece of flat plate would be marked out as drawn below ready for forming in the rolls.

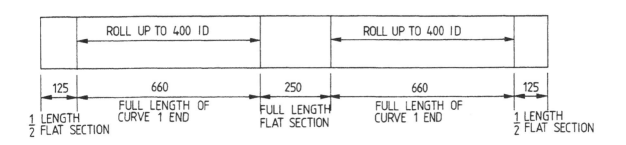

ROLL UP TO 400 ID

ROLL UP TO 400 ID

| 125 | 660 | 250 | 660 | 125 |

½ LENGTH FLAT SECTION

FULL LENGTH OF CURVE 1 END

FULL LENGTH FLAT SECTION

FULL LENGTH OF CURVE 1 END

½ LENGTH FLAT SECTION

Example 3

A chute plate has the sectional shape shown below. Calculate the length of plate required to form this plate.

Material: 8 mm plate

$$MD = (2 \times IR) + T$$
$$= (2 \times 225) + 8$$
$$= 450 + 8$$
$$= 458 \text{ mm}$$

Length = straight sections + curved sections

$$= [2 \times (150 - T)] + [2 \times (130 - T)]$$
$$+ 360 + \left(\frac{MD \times \pi}{2} \right)$$

$$= (2 \times 142) + (2 \times 122) + 360$$
$$+ \left(\frac{458 \times 3.1416}{2} \right)$$

$$= 284 + 244 + 360 + \frac{1438.8528}{2}$$

$$= 284 + 244 + 360 + 719.4264$$

$$= 1607.4264$$
   say, 1608 mm

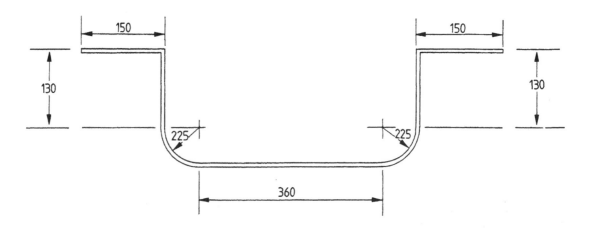

The plate would be marked out as shown below ready to form in the rolls and press.

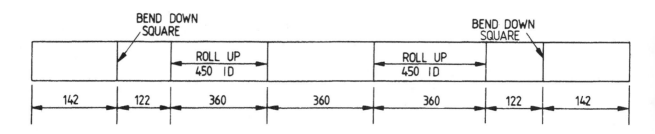

Example 4

Calculate the length of flat required to bend the support bracket drawn below.

Material: 100 × 5 flat bar

Length = base + 2 verticals + 2 tops
= (300 − 2 T) + [2 × (180 − 2 T)] +
  [2 × (50 − T)]
= 290 + (2 × 170) + (2 × 45)
= 290 + 340 + 90
= 720 mm

The flat bar would be marked out as drawn below ready to press up.

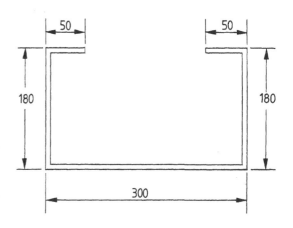

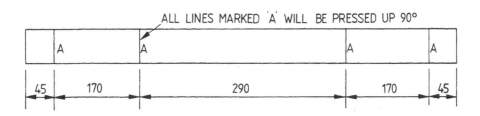

ALL LINES MARKED 'A' WILL BE PRESSED UP 90°

There are many more combinations of shapes and sizes, but from the preceding examples you should gain enough understanding to be able to calculate the correct lengths of metal to be formed.

Exercise 1

(a) Calculate the length of material required to form the pipe bracket drawn below.

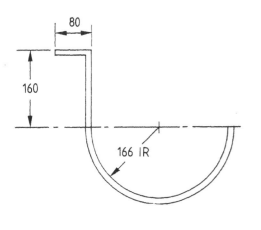

Material: 100 × 10 flat bar

Length = straight sections + curved sections

=

=

=

=

(b) Mark out the flat bar ready to be formed in the rolls and the press.

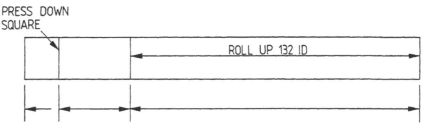

PRESS DOWN SQUARE

ROLL UP 132 ID

Exercise 2

(a) Calculate the length of material to form the opening stiffener drawn below.

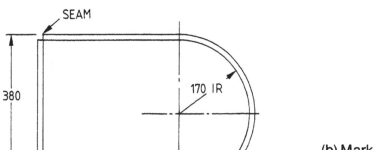

Material: 200 × 20 flat bar

Length = straight sections + curved section

=

=

=

=

(b) Mark out the flat bar ready to be formed in the roll and the press.

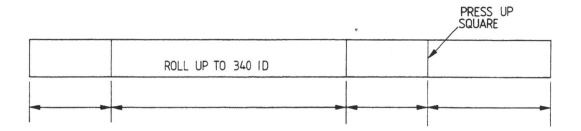

- -

Exercise 3

(a) Calculate the length of plate required to form the chute as drawn below.

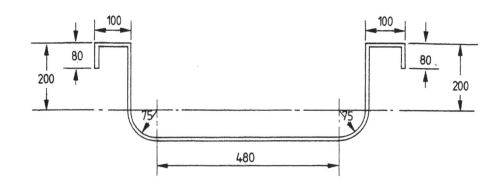

Material: 6 mm plate

(b) Mark the chute plate ready for pressing and rolling.

Length = straight sections + curved sections

=

=

=

=

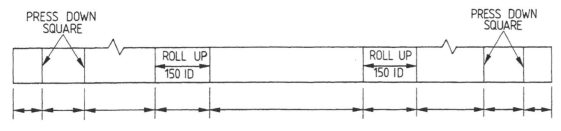

PRESS DOWN SQUARE

PRESS DOWN SQUARE

ROLL UP
150 ID

ROLL UP
150 ID

Exercise 4

(a) Calculate the length of plate required to form the support bracket as drawn below.

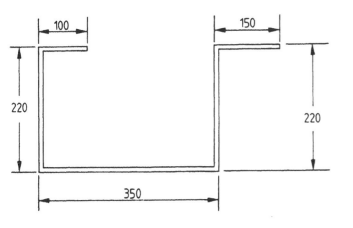

100

150

220

220

350

Material: 150 × 12 flat bar

Length =

=

=

=

=

(b) Mark out the flat bar ready for forming in the press.

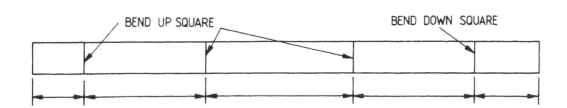

BEND UP SQUARE

BEND DOWN SQUARE

6.9 Size of plates for dished ends

Dished ends for pressure vessels

Dished ends fall into two categories:

1. When an end plate is formed to a half sphere it is referred to as a *hemispherical* end.
2. When an end plate is formed to a partial sphere it is referred to as either an *ellipsoidal* or a *torispherical* end.

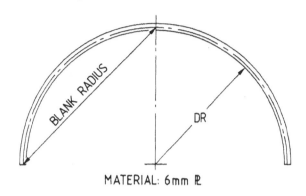

MATERIAL: 6mm ℞

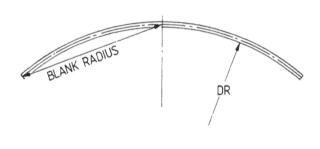

Determination of blank diameter for dished ends

The diameter of a flat blank required to form a dished end may be found by either calculation or from a full-sized layout. The calculation varies according to forming process, so the layout method will be dealt with in detail.

The layout method requires the putting down of a full-sized sectional elevation of the end to be formed, and in order to complete this layout the following information is required:

1. Outside diameter of the shell to which the end is to be attached.
2. Dishing radius.
3. Knuckle or corner radius.
4. Plate thickness.
5. Type of connection between end and shell (butt or lap).

Layout of end

1. Draw baseline *AB* and construct verti_ centre line.
2. Mark points 1,1, the ID of the dished e_ apart, and equidistant from the centre lin_
3. Erect perpendiculars on points 1,1 a_ mark up these the flange lengths to loca_ points 2, 2.
4. Join points 2 and 2 and locate points 3, 3 marking in from points 2, the knuckle radi_
5. Locate centre point 4 of dishing radius _ subtracting knuckle radius from dishi_ radius, setting the trammels on that dime_ sion and striking arcs from points 3 to c_ the vertical centre line.
6. Locate tangent point between knuckle ar_ dishing radii by projecting a line from poi_ 4 through points 3.
7. Scribe in knuckle and dishing radii and ac_ plate and mean plate thickness lines.

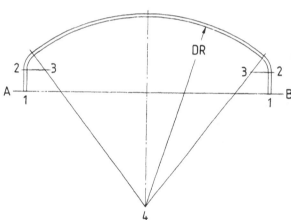

MATERIAL: 6mm ℞

Blank diameter

In the layout of the dished end shown below, the points A, B, C and D are located on the mean line of the plate thickness, AB representing the straight flange, BC and CD the chordal lengths of the knuckle and half dish respectively. The sum of AB, BC and CD will give the radius of the blank plate required to form the dished flanged end.

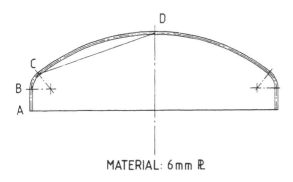

MATERIAL: 6mm ℞

Calculation

$$d = (AB + BC + CD + \text{plate thickness}) \times 2$$
$$= (65 + 59 + 360 + 6) \times 2$$
$$= 481 \times 2$$
$$= 962 \text{ mm}$$

Note To the theoretical radius must be added a trimming allowance of approx. 1 × plate thickness to account for any slight misalignment during forming. Excessive trimming allowances will promote wrinkling during forming.

Alternative method for blank diameter

In the alternative method shown below, the mean thickness line of the flange and knuckle radius on the left-hand side is divided into a number of equal divisions, AB, BC, etc. On the right-hand side, the mean dishing radius is extended in order that the dimensions AB, BC, etc., can be added to produce HA.

The chordal distance HA is the theoretical radius of plate required for the dished flanged end.

When dividing the line DG, the spacings must be such that there is no appreciable difference between the lengths of arc and their chords.

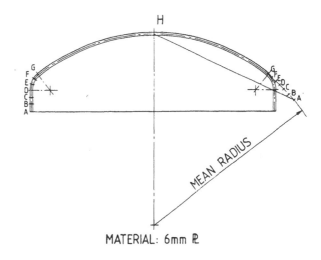

MATERIAL: 6mm ℞

Calculation

$$d = (HA + \text{plate thickness}) \times 2$$
$$= (460 + 6) \times 2$$
$$= 466 \times 2$$
$$= 932 \text{ mm} + \text{trimming allowance}$$

6.10 Calculations for conical work

By the use of calculations, all right cones and frustums of right cones can be developed without the use of an elevation. Shown below are a series of calculations that will enable you to develop right cones without the elevation being drawn.

Frustum of right cones
Calculations for layout and development
Formulae

$$\text{Apex height: } H = \frac{D \times h}{(D - d)}$$

$$\text{Slant height: } SH = \sqrt{H^2 + R^2}$$

$$\text{Small circumference: } c = \pi d$$

$$\text{Large circumference: } C = \pi D$$

$$\text{Slant height of frustum: } sh = \sqrt{h^2 + (R - r)^2}$$

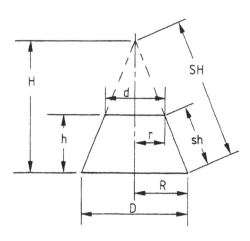

Example

Calculate the apex height, slant height, small and large circumference of a frustum of a right cone where:

D = 2700 mm
d = 950 mm
h = 1290 mm

$$H = \frac{D \times h}{(D - d)}$$
$$= \frac{2700 \times 1290}{(2700 - 950)}$$
$$= \frac{3\,483\,000}{1750}$$
$$= 1990.285 \text{ mm}$$

$$sh = \sqrt{h^2 + (R - r)^2}$$
$$= \sqrt{1290^2 + (1350 - 475)^2}$$
$$= \sqrt{1290^2 + 875^2}$$
$$= \sqrt{1\,664\,100 + 765\,625}$$
$$= \sqrt{2\,429\,725}$$
$$= 1558.7575 \text{ mm}$$
$$\text{say, } 1559 \text{ mm}$$

$$SH = \sqrt{H^2 + R^2}$$
$$= \sqrt{3\,961\,234 + 1\,822\,500}$$
$$= \sqrt{5\,783\,734}$$
$$= 2404.939 \text{ mm}$$

$$c = \pi d$$
$$= 3.1416 \times 950$$
$$= 2984.52 \text{ mm}$$

$$C = \pi D$$
$$= 3.1416 \times 2700$$
$$= 8482.32 \text{ mm}$$

Calculate the apex height, small and large circumference and slant height of a frustum of a right cone.

D = 2900 mm
d = 850 mm
h = 1390 mm

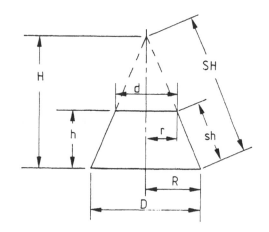

$$H = \frac{D \times h}{(D - d)}$$

$$=$$

$$=$$

$$= \qquad mm$$

$$c = \pi d$$

$$=$$

$$= \qquad mm$$

$$C = \pi D$$

$$=$$

$$= \qquad mm$$

$$SH = \sqrt{H^2 + R^2}$$

$$= \sqrt{}$$

$$= \sqrt{}$$

$$= \qquad mm$$

say, mm

$$sh = \sqrt{h^2 + \left(\frac{D - d}{2}\right)^2}$$

$$= \sqrt{}$$

$$= \sqrt{}$$

$$= \sqrt{}$$

$$= \sqrt{}$$

$$= \qquad mm$$

say, mm

Frustum of oblique cone

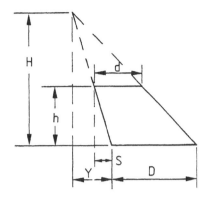

Apex height $= H$

$$H = \frac{D \times h}{(D - d)}$$

Horizontal distance $= Y$

$$Y = \frac{D \times S}{(D - d)}$$

For developing pattern of cone (verse sine)

where: D = mean diameter of base
        d = mean diameter of top
        h = perpendicular height of frustum
       H = height of apex above base
    SH = slant height of right cone
           i.e., radius for development of
           large end

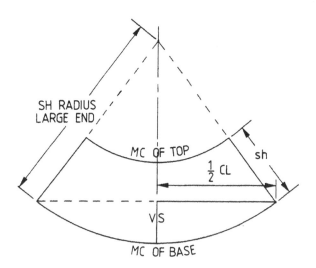

Verse sine

$$VS = \frac{D^2 \times 1.2337}{SH}$$

Half chordal length

$$\tfrac{1}{2}\,CL = \sqrt{(2 \times SH - VS) \times VS}$$

Slant height of frustum

$$sh = \sqrt{h^2 + \left(\frac{D - d^2}{2}\right)}$$

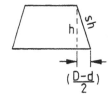

MC top = $\pi \times$ md
MC base = $\pi \times$ MD

Note 1. Diameters are mean diameters. Use millimetres in calculations.
      2. This method should only be used on slow tapering cones.

Example

By calculations, determine the desired information for the development of the following cone by the verse sine method.

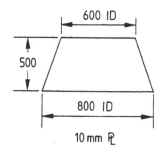

md of top = ID + T
          = 600 + 10
          = 610 mm
MD of base = ID + T
          = 800 + 10
          = 810 mm

$$H = \frac{\dot{D} \times h}{(D - d)}$$
$$= \frac{810 \times 500}{(810 - 610)}$$
$$= \frac{810 \times 500}{200}$$
H = 2025 mm

$$SH = \sqrt{H^2 + R^2}$$
$$= \sqrt{2025^2 + 405^2}$$
$$= \sqrt{4\,100\,625 + 164\,025}$$
$$= \sqrt{4\,264\,650}$$
$$= 2065.103$$
SH = 2065 mm

$$VS = \frac{D^2 \times 1.2337}{SH}$$
$$= \frac{810^2 \times 1.2337}{2065}$$
$$= 391.976$$
VS = 392 mm

$$\tfrac{1}{2}\,CL = \sqrt{(2 \times SH - VS) \times VS}$$
$$= \sqrt{(2 \times 2065 - 392) \times 392}$$
$$= \sqrt{3738 \times 392}$$
$$= \sqrt{1\,465\,296}$$
$$= 1210.49$$
$\tfrac{1}{2}\,CL$ = 1210 mm

$$sh = \sqrt{h^2 + \left(\frac{D-d}{2}\right)^2}$$

$$= \sqrt{500^2 + \left(\frac{810-610}{2}\right)^2}$$

$$= \sqrt{500^2 + 100^2}$$
$$= \sqrt{250\,000 + 10\,000}$$
$$= \sqrt{260\,000}$$
$$= 509.9$$
$$sh = 510 \text{ mm}$$

$$\begin{aligned}
\text{MC top} &= \pi \times md \\
&= \pi \times 610 \\
&= 1916
\end{aligned}$$

$$\tfrac{1}{2} \text{ MC top} = 958 \text{ mm}$$

$$\begin{aligned}
\text{MC base} &= \pi \times MD \\
&= \pi \times 810 \\
&= 2545
\end{aligned}$$

$$\tfrac{1}{2} \text{ MC base} = 1272.5 \text{ mm}$$

The following formula can be used to calculate any diameter parallel to the base.

To determine diameter at any diameter parallel to base:

$$d1 = \frac{H1 \times D}{H}$$

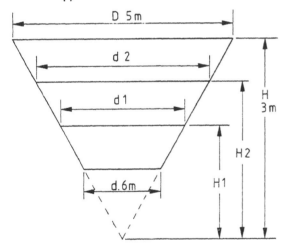

Example

To find diameters d1 and d2:

$$\begin{aligned}
d1 &= \frac{H1 \times D}{H} \\
&= \frac{1.409 \times 5}{3.409} \\
&= 2.067 \text{ m}
\end{aligned}$$

$$\begin{aligned}
d2 &= \frac{H2 \times D}{H} \\
&= \frac{2.409 \times 5}{3.409} \\
&= 3.533 \text{ m}
\end{aligned}$$

$$\begin{aligned}
H &= \frac{D \times h}{D - d} \\
&= \frac{5 \times 3}{5 - 0.6} \\
&= 3.409 \text{ m}
\end{aligned}$$

Conical tanks

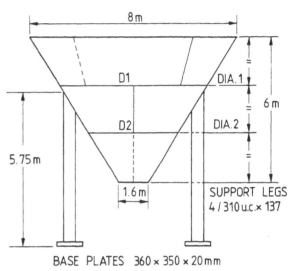

BASE PLATES 360 × 350 × 20 mm

Exercise

From the above sketch calculate apex height, diameter 1 and diameter 2.

$$H = \frac{D \times h}{D - d}$$

$$=$$

$$= \qquad \text{m}$$

$$D1 = \frac{H1 \times D}{H}$$

$$=$$

$$= \qquad \text{m}$$

$$D2 = \frac{H2 \times D}{H}$$

$$=$$

$$= \qquad \text{m}$$

6.11 Pythagoras' Theorem

Calculating the slant height of a right-angled triangle using Pythagoras' Theorem can often eliminate the need for a layout when developing the size of slant heights.

Pythagoras' Theorem

In a right-angled triangle, the square of the hypotenuse equals the sum of the squares on the other two sides.

$$C^2 = A^2 + B^2$$

By transposition:

$$A^2 = C^2 - B^2$$
and
$$B^2 = C^2 - A^2$$

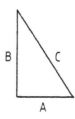

Therefore:
$$C = \sqrt{A^2 + B^2}$$
$$A = \sqrt{C^2 - B^2}$$
$$B = \sqrt{C^2 - A^2}$$

Pythagoras' Theorem where A and B are known.

$$C^2 = A^2 + B^2$$
$$C = \sqrt{A^2 + B^2}$$

Example 1

$A = 3\,m, \ B = 4\,m, \ C = ?$

$$
\begin{aligned}
C &= \sqrt{A^2 + B^2} \\
&= \sqrt{3^2 + 4^2} \\
&= \sqrt{9 + 16} \\
&= \sqrt{25} \\
&= 5\,m
\end{aligned}
$$

From any right-angled triangle, where lengths B and C are known, length A can be calculated. Also if lengths C and A are known, then length B can be calculated.

Example 2

Calculate length of base.
$C = 5\,m, \ B = 4\,m, \ A = ?$

$$
\begin{aligned}
A^2 &= C^2 - B^2 \\
A &= \sqrt{C^2 - B^2} \\
&= \sqrt{5^2 - 4^2} \\
&= \sqrt{25 - 16} \\
&= \sqrt{9} \\
&= 3\,m
\end{aligned}
$$

Example 3

Calculate length of vertical height.
$C = 5\,m, \ A = 3\,m, \ B = ?$

$$
\begin{aligned}
B^2 &= C^2 - A^2 \\
B &= \sqrt{C^2 - A^2} \\
&= \sqrt{5^2 - 3^2} \\
&= \sqrt{25 - 9} \\
&= \sqrt{16} \\
&= 4\,m
\end{aligned}
$$

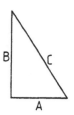

Applications for Pythagoras' Theorem will appear in these other sections of the book.

1. Checking of diagonals — Chapter 10
2. Conical work — Chapter 11
3. Triangulation — Chapter 13
4. Hoppers — Chapter 14

Exercise 1

Calculate length of hypotenuse.

$$
\begin{aligned}
C &= \sqrt{A^2 + B^2} \\
&= \sqrt{\quad\quad} \\
&= \sqrt{\quad\quad} \\
&= \sqrt{\quad\quad} \\
&= \quad\quad mm
\end{aligned}
$$

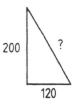

Exercise 2

Calculate length of base.

$$
\begin{aligned}
A &= \sqrt{C^2 - B^2} \\
&= \sqrt{\quad\quad} \\
&= \sqrt{\quad\quad} \\
&= \sqrt{\quad\quad} \\
&= \quad\quad mm
\end{aligned}
$$

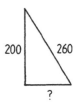

Exercise 3
Calculate length of vertical height.

$$B = \sqrt{C^2 - A^2}$$
$$= \sqrt{\rule{3cm}{0.4pt}}$$
$$= \sqrt{\rule{3cm}{0.4pt}}$$
$$= \sqrt{\rule{3cm}{0.4pt}}$$
$$= \quad \text{mm}$$

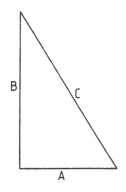

?   250

80

Further use of Pythagoras' Theorem using square root tables

By applying these formulae to any right-angled triangle, we can find the length of a side given the lengths of the other two sides.

Example

An angle frame is in the form of a right-angled triangle and has a base length of 9000 mm and a perpendicular height of 7000 mm. What is the length of the third side?

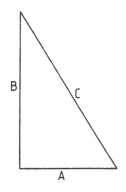

B   C

A

Base length, $B = 9000$ mm $= 9$ m
Perp. height, $A = 7000$ mm $= 7$ m
   Third side, $C = \sqrt{A^2 + B^2}$
$$= \sqrt{9^2 + 7^2}$$
$$= \sqrt{81 + 49}$$
$$= \sqrt{130}$$
from tables $= 11.4018$ m

Exercise 1
Calculate the length of bracing required to tie two sides of a hopper together.

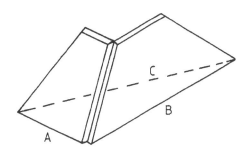

C

B

A

$B = 5000$ mm $= 5.0$ m
$A = 3000$ mm $= 3.0$ m

$$C = \sqrt{B^2 + A^2}$$
$$= \sqrt{\rule{3cm}{0.4pt}}$$
$$= \sqrt{\rule{3cm}{0.4pt}}$$
$$= \sqrt{\rule{3cm}{0.4pt}}$$
from tables $=$ m

Exercise 2
Calculate length of diagonal measurement from sketch using square root tables.

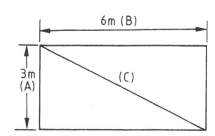

6m (B)

3m (A)

(C)

$$C = \sqrt{A^2 + B^2}$$
$$= \sqrt{\rule{3cm}{0.4pt}}$$
$$= \sqrt{\rule{3cm}{0.4pt}}$$
$$= \sqrt{\rule{3cm}{0.4pt}}$$
from tables $=$ m

Another method for finding the length of hypotenuse of a right-angled triangle.

Use of maths tables to find the hypotenuse of a triangle

Example

To find the length of the hypotenuse of a right-angled triangle with a base of 450 mm and a height of 1275 mm.

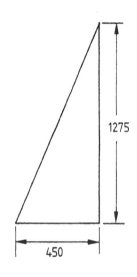

Steps

1. Divide the base into the height.
$$\frac{1275}{450} = 2.8333$$
2. Look up the tangent tables to find the closest number to 2.8333 = 70.34.
3. Turn to natural secants and find the coefficient for 70.34 = 3.0057.
4. Multiply this coefficient (3.0057) by the base length (450) to obtain the hypotenuse = 1352.56 mm.

Exercises

Find the length of the hypotenuse of the following right-angled triangles.

1. Base = 425, height = 875.

Step 1 =
Step 2 =
Step 3 =
Step 4 =

2. Base = 920, height = 360.

Step 1 =
Step 2 =
Step 3 =
Step 4 =

3. Base = 325, height = 375.

Step 1 =
Step 2 =
Step 3 =
Step 4 =

4. Base = 225, height = 875.

Step 1 =
Step 2 =
Step 3 =
Step 4 =

6.12 Calculating lengths for internal and external angle rings

Internal angle ring

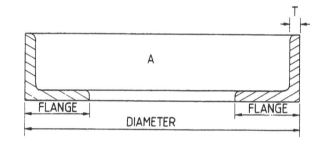

External angle ring

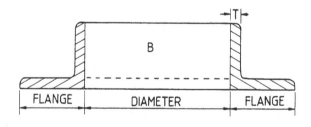

D = Heel diameter
T = Flange thickness
F = Flange width (horizontal)

Internal angle ring is one where the toe of the angle looks into centre A. The diameter is always measured from heel to heel.

External angle ring is one where the toe of the horizontal flange looks away from centre B.

Length of angle

Internal

$$\text{Length} = 3.1416 \times \left(D - T - \frac{F}{3}\right)$$

External

$$\text{Length} = 3.1416 \times \left(D + T + \frac{F}{3}\right) + \frac{2}{3}F \text{ allowance}$$

Example 1

(a) Calculate the length of angle required to form the external angle drawn below.

Material: 64 × 64 × 6 angle

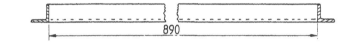

$$\begin{aligned}
\text{Length} &= 3.1416 \times \left(D + T + \frac{F}{3}\right) \\
&= 3.1416 \times \left(890 + 6 + \frac{64}{3}\right) \\
&= 3.1416 \times (896 + 21.3) \\
&= 3.1416 \times 917.3 \\
&= 2881.789
\end{aligned}$$

say, 2882 mm + 42.6 mm allowance (⅓ flange each end)

(b) Mark out the angle ready to be cut and formed.

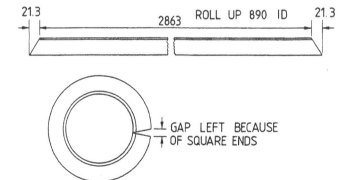

External angle ring after forming shows the need for an allowance for ends to be cut out of square as shown on the sketch above.

Example 2

(a) Calculate the length of angle required to form the internal angle ring as drawn below.

Material: 44 × 44 × 5 angle

$$\begin{aligned}
\text{Length} &= 3.1416 \times \left(D - T - \frac{F}{3}\right) \\
&= 3.1416 \times \left(520 - 5 - \frac{44}{3}\right) \\
&= 3.1416 \times (515 - 14.6) \\
&= 3.1416 \times 500.4 \\
&= 1572.0566 \text{ mm}
\end{aligned}$$

say, 1572 mm

(b) Mark out the angle ready to be cut and formed.

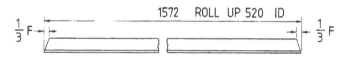

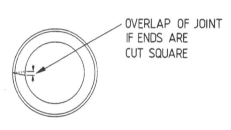

Internal angle ring after forming shows the need for an allowance for ends to be cut out of square as shown on sketch above.

Exercise 1

Calculate the length of angle required to form the external angle ring as drawn below.

Material: 75 × 75 × 10 angle

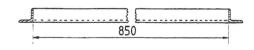

850

$$\text{Length} = 3.1416 \times \left(D + T + \frac{F}{3}\right)$$

$$=$$
$$=$$
$$=$$
$$=$$

say,   mm +   mm allowance

Exercise 2

Calculate the length of angle required to form the internal angle ring drawn below.

Material: 64 × 64 × 8 angle

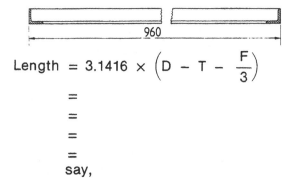

960

$$\text{Length} = 3.1416 \times \left(D - T - \frac{F}{3}\right)$$

$$=$$
$$=$$
$$=$$
$$=$$

say,

6.13 Calculating fine angles from radian and tangent tables

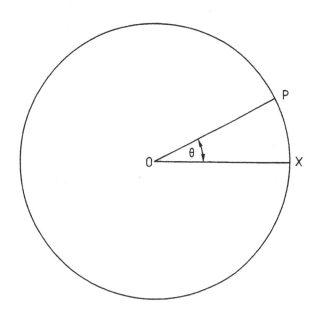

When the ray *OP* starting from *OX* travels (counterclockwise) through a whole revolution to again reach *OX*, we say the angle described is 360° or 2π radians = 6.283 rads.

Therefore: If we know the angle (θ), then point *P* can be found on the circumference by multiplying the coefficient of the angle by the radius *OX*.

Note The coefficient of the angle is read from the mathematical tables "Degrees and Radians".

Example 1

Position a bracket at *P* on a cylinder with an outside diameter of 4.500 m, so that its position is 27°30′ above the horizontal centre line.

1. From degrees and radians tables select the coefficient for 27°30′ = 0.4800.
2. Multiply this coefficient (0.4800) by the outside radius, 0.4800 × 2.250 m = 1.0800 m.
3. Measure 1.0800 m from the horizontal centre line around the outside circumference to obtain the required position.

Example 2

Position of holes on layout using tangent tables. Hole position marked from an existing hole by calculating height O from given angle, 23°18′.

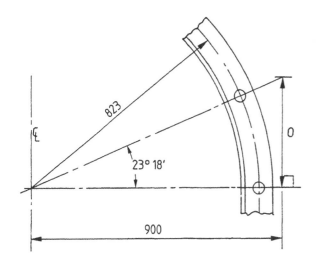

1. Select any convenient length greater than 823 mm (say 900 mm).
2. From tables select the coefficient for 23°18′ = 0.4307.
3. To find height O, multiply the coefficient (0.4307) by the selected length (900) = 387.63 mm.
4. Construct the right-angled triangle to give the required angle and to locate hole.

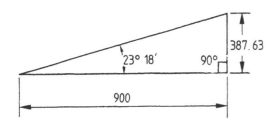

Example 3

To construct angles of any size measured around a curved surface, using degree and radian tables.

Position a hole on an angle ring with a PCD of 4.500 m, so that its position is 27°30′ above the horizontal centre line.

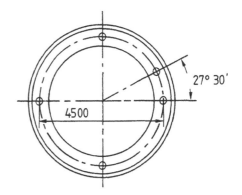

1. From degree and radian tables select the coefficient for 27°30′ = 0.4800.
2. Multiply this coefficient (0.4800) by the radius (radius = 2.250 m), 0.4800 × 2.250 m = 1.080 m.
3. Measure 1.080 m from the horizontal centre line around the PC diameter to obtain the required position.

Exercise 1

Calculate the position of the butt joint of a 5.200 m heel diameter cylindrical angle ring if located 30°19′ from the horizontal centre line.

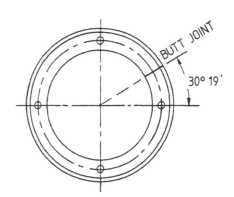

1. Select the coefficient from radian tables for 30°19′ =

2. Multiply $\qquad \times \dfrac{\text{diam.}}{2}$

   =

   =

   = measured around the ring

Exercise 2

Calculate the position of a hole on an angle ring with a PCD of 2.38 m if the hole is 15°40′ from the horizontal.

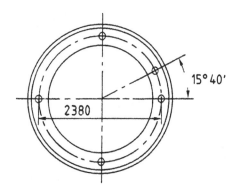

1. Select the coefficient from the radian tables for 15°40′ =

2. Multiply $\qquad \times \dfrac{\text{diam.}}{2}$

$\qquad\qquad =$

$\qquad\qquad =$

$\qquad\qquad =$ measured around PCD

Exercise 3

Calculate the position of centre line of branch pipe which is positioned 24° from northern centre line of tank measuring 3.048 m outside diameter.

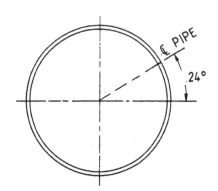

1. Select the coefficient from the radian tables for 24° =

2. Multiply $\qquad \times \dfrac{\text{diam.}}{2}$

$\qquad\qquad =$

$\qquad\qquad =$

$\qquad\qquad =$ measured around outside of tank

Laying out angles other than right angles

Using natural tangent tables

Example

Construct an angle of 23°18′ accurate over a length of 823 mm.

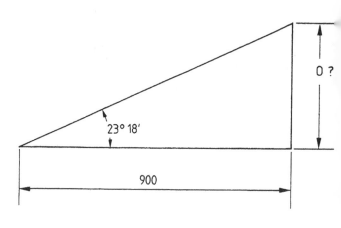

1. Select any length greater than 823 mm (say 900).
2. From natural tangents tables select the coefficient for 23°18′ = 0.4307.
3. To find the height O, multiply the coefficient (0.4307) by the selected length (900) = 387.63 mm.
4. Construct the right-angled triangle to give the required angle.

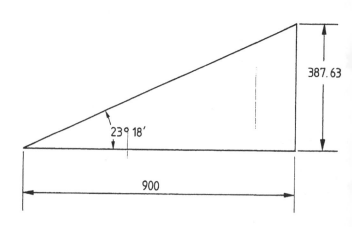

Using natural tangents tables, calculate the height of the following angles

Exercise 1

Angle 30° with a base length of 900 mm

        30° angle =
            base = 900
          height =
                 =

Exercise 2

Angle 40°24′ with a base length of 1000 mm

    40°24′ angle =
            base = 1000
          height =
                 =

Exercise 3

Angle 35°18′ with a base length of 800 mm

    35°18′ angle =
            base = 800
          height =
                 =

Exercise 4

Angle 71°6′ with a base length of 3200 m

     71°6′ angle =
            base = 3200 m
          height =
                 =

NOTES

7

Elementary print reading

Engineering drawings are normally intended to indicate the shape and size of an object. However, all objects have three dimensions, length, breadth and depth, and the problem of representing these on a drawing as well as conveying an impression of the shape to the tradesman is overcome by the use of a technique called orthogonal projection.

In order to interpret orthogonal drawings correctly, it is very important for the boilermaker marker-off to be able to distinguish between first and third angle projections. Any misinterpretation, even of a relatively simple component, will result in some detail being fabricated in the opposite hand.

All drawings in this book are based on third angle orthogonal projection. The next two pages will help you understand this type of projection.

It is essential that drawings made by the third angle method be identified, preferably by the use of the standards symbol illustrated below or by the words "Third Angle Projection" printed in a conspicuous place on the drawing, such as the title block.

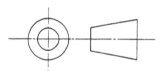

PROJECTION SYMBOL

Third angle projection symbol

Also included in this chapter are explanations and examples of the types of lines you will come across on the various drawings (pages 128–9) and brief explanation, examples and exercises on metric scales as used on scale drawings (pages 130–1).

Note

Do not measure from the diagrams except in exercises, because many figures are indication diagrams only and are not drawn to scale. All measurements on the diagrams are in millimetres unless stated otherwise.

Drawn below is a practical example of third angle projection. Note that the respective view is projected to the same side that you look at (mirror image).

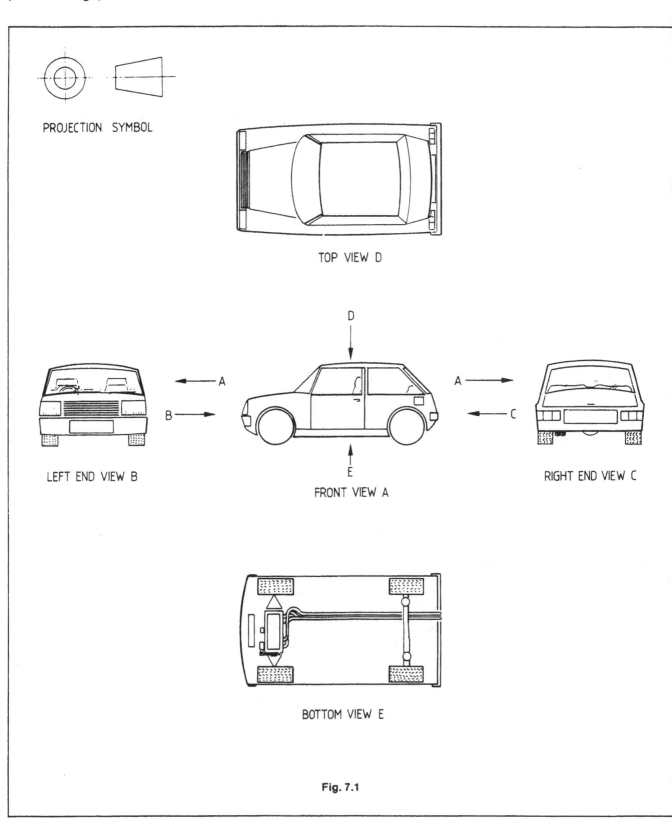

PROJECTION SYMBOL

TOP VIEW D

LEFT END VIEW B

FRONT VIEW A

RIGHT END VIEW C

BOTTOM VIEW E

Fig. 7.1

Drawn below is a practical example of first angle projection. Note that the respective view is projected to the opposite side that you are looking at.

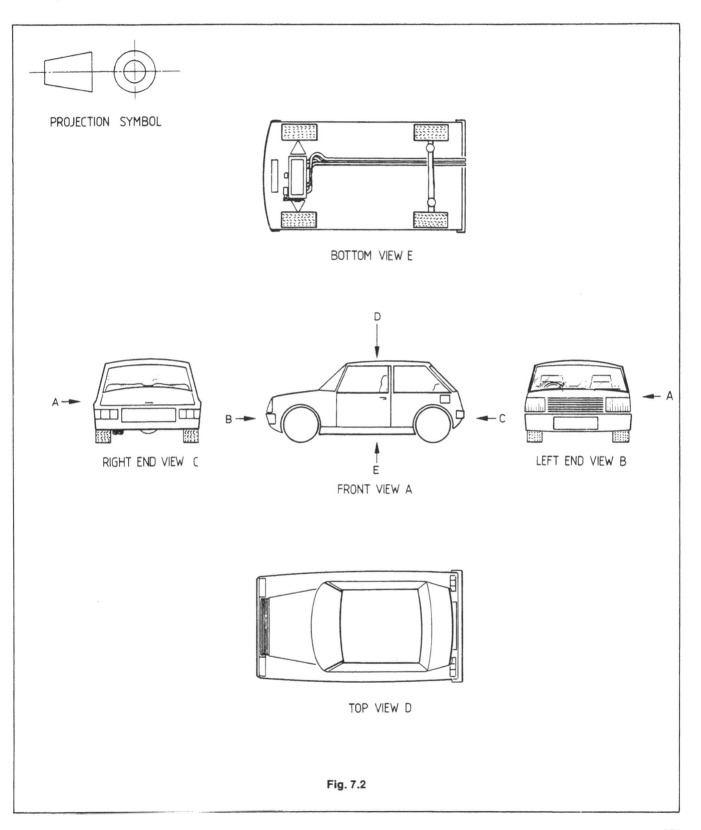

PROJECTION SYMBOL

BOTTOM VIEW E

RIGHT END VIEW C

FRONT VIEW A

LEFT END VIEW B

TOP VIEW D

Fig. 7.2

Orthogonal projection

What is pictorial drawing?

A pictorial drawing is a drawing of an object in three dimensions showing three of its sides. The type used in this book is isometric drawing.

What is orthogonal drawing?

An orthogonal drawing is a drawing of a number of views of an object, each view showing the shape and size of one side of the object. Each view is drawn in such a position that information can be projected from one view to another.

Figure 7.3 is a pictorial drawing of a metal box showing three sides only.

Figure 7.4 shows the box opened out, showing all six sides.

In Figure 7.5 the sides have been separated so that the necessary dimensions or sizes of the box can be shown.

Note The rear side has been omitted.

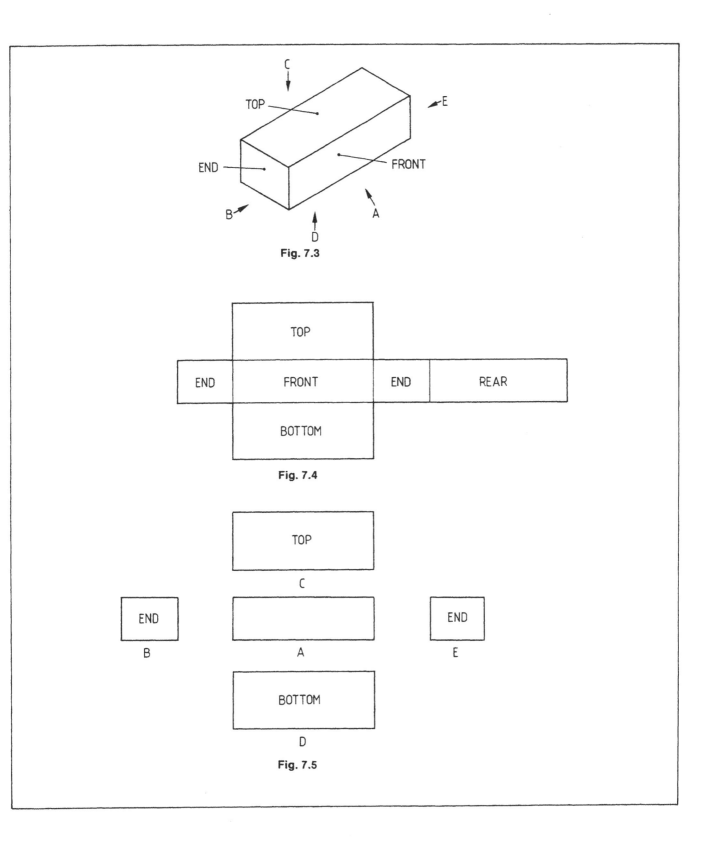

Fig. 7.3

Fig. 7.4

Fig. 7.5

Third angle projection is the recommended system of projection and is used for all exercises in this book.

The front view is a side view and is the first view to be drawn. It is from this view that most other views are projected. The views must be in correct alignment with the front view as in Figure 7.7.

The three views, A, B, C would be sufficient for the simple object shown in Figure 7.6. The views D and E would only be drawn if special features were to be shown on these faces.

A view from the **left** is drawn on the **left**.
A view from the **right** is drawn on the **right**.
A view from the **top** is drawn on the **top**.
A view from **underneath** (known as a bottom view) is drawn **underneath**.

Note Rear view not shown.

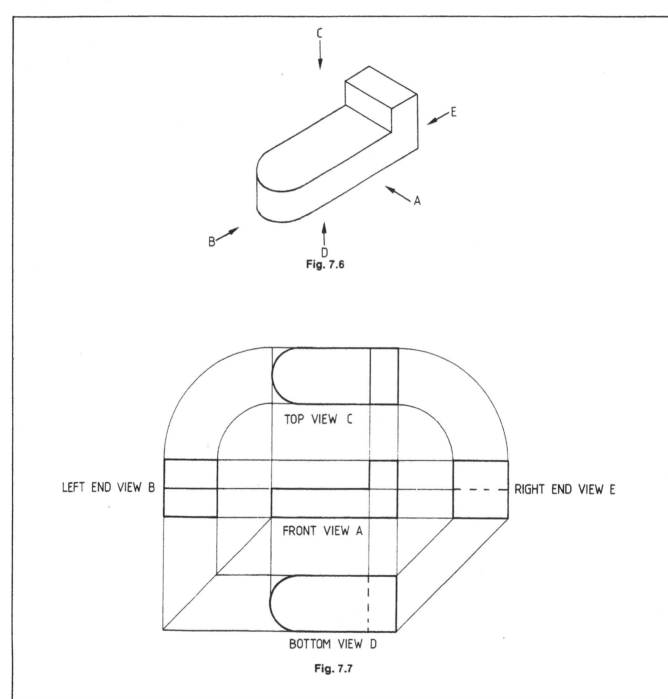

Fig. 7.6

LEFT END VIEW B

TOP VIEW C

FRONT VIEW A

RIGHT END VIEW E

BOTTOM VIEW D

Fig. 7.7

Third angle orthogonal projection

Standards

Third angle orthogonal projection is the system of projection recommended by the Australian Standards Association for drawing practice. (Code No. AS1100).

What is third angle projection?

Third angle projection is the name given to the arrangement of the views relative to the *front view* (see Fig. 7.9).

Note Each view is given a special name.

Rule for third angle projection (reference to Fig. 7.9)
1. A view from the **left** of the front view is drawn on the **left** (left end view *B*).
2. A view from the **right** of the front view is drawn on the **right** (right end view *E*).
3. A view from the **top** of the front view is drawn on the **top** (top view *C*).
4. A view from the **underside** of the front view is drawn on the **underside** (bottom view *D*).

Note Since this is a simple object, only three views are necessary to show its size and shape (views *A*, *B*, and *C*).

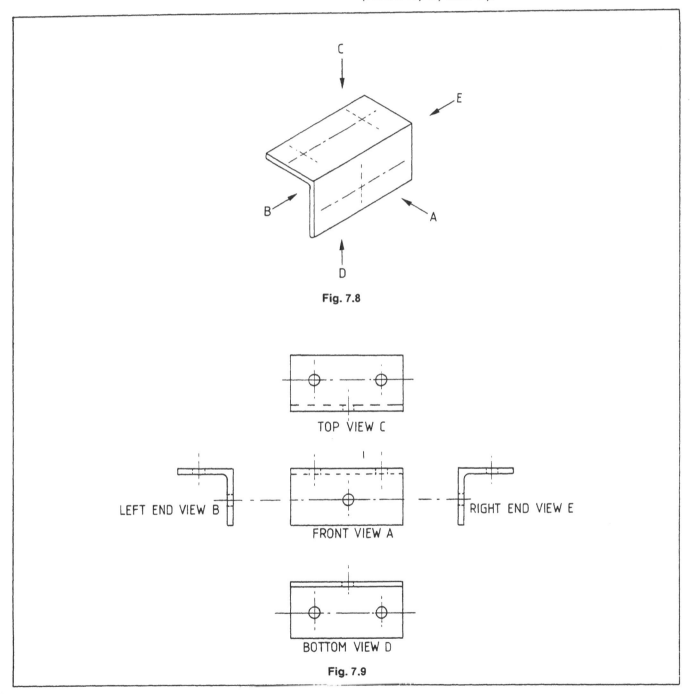

Fig. 7.8

Fig. 7.9

Practical examples

Drawn below are four jobs showing the application of third angle projection. Without all the views shown, the correct interpretation of shape and size would not be possible.

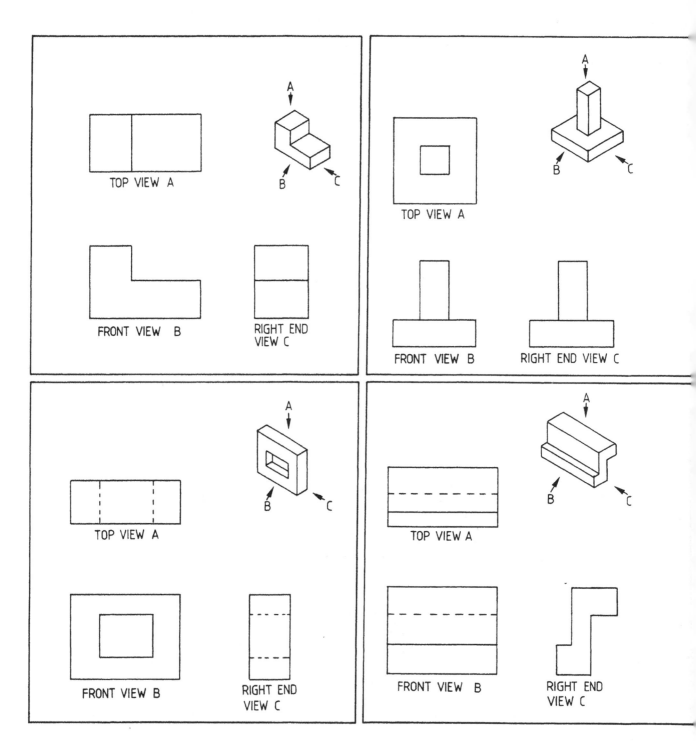

TOP VIEW A

FRONT VIEW B

RIGHT END VIEW C

TOP VIEW A

FRONT VIEW B

RIGHT END VIEW C

TOP VIEW A

FRONT VIEW B

RIGHT END VIEW C

TOP VIEW A

FRONT VIEW B

RIGHT END VIEW C

Below is a fabricated bearing block bracket
drawn in third angle projection.

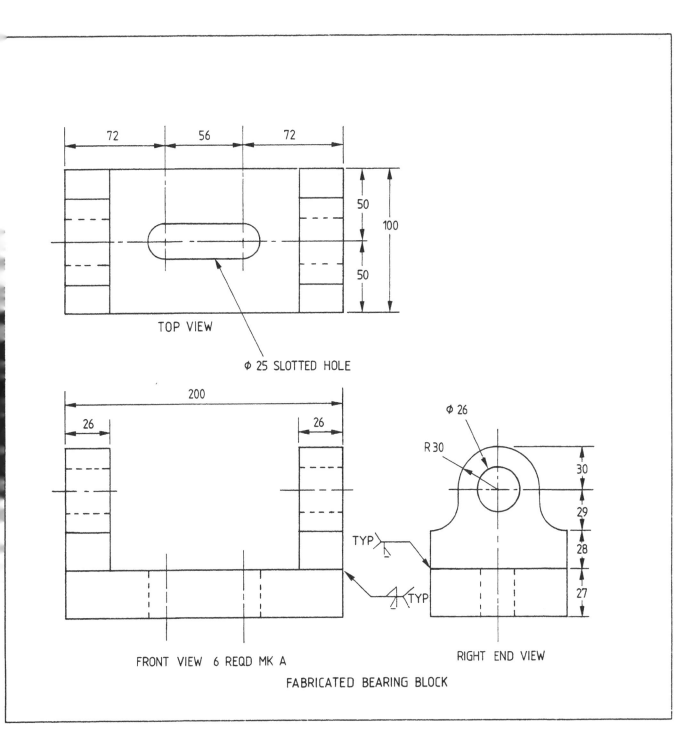

FABRICATED BEARING BLOCK

Types of lines used on metal fabrication drawings

Types of lines	Examples of lines
1. Outlines (a) Visible outlines	———————— thick
(b) Hidden outlines	– – – – medium – – – – –
2. Centres (a) Centre lines	thin
(b) Pitch lines	
(c) Pitch circles	
3. Dimensions (a) Dimension lines	thin
(b) Extension lines	
(c) Leader lines	
4. Breaks (a) Short break lines	thin
(b) Long break lines	
(c) 'S' break lines	
5. Sections (a) Hatching lines	thin
(b) Cutting place lines	thick thin thick

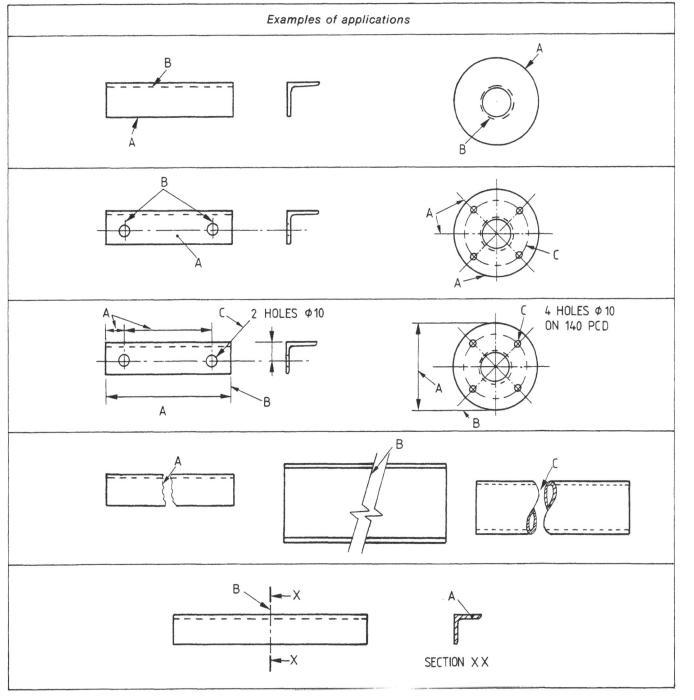

Examples of applications

2 HOLES φ10

4 HOLES φ10 ON 140 PCD

SECTION X X

Scale drawing

To enable drawing sheets to be made a convenient size, engineering details are often drawn to a reduced or enlarged scale. Large details, structures and machine parts are drawn smaller than actual size, while small parts, such as parts of instruments, etc., are drawn larger than their actual size. Figure 7.10 is a full size drawing of an angle section. Figure 7.11 and Figure 7.12 are drawings of the same angle made to the scale of half (1:2) and one-fifth (1:5) of full size respectively. Comparing Figures 7.10, 7.11 and 7.12 it will be seen that the ratio of length to length is constant thus making each drawing identical in shape.

Scales recommended for use on engineering drawings

The following scales are recommended in the Australian Standard Drawing Practice AS1100, Part 7–1972, for use on engineering drawings.

Metric scales

Full size 1:1.
Reducing scales 1:2, 1:5, 1:10, 1:20, 1:50, 1:10
1:200.
Enlargements 2:1, 5:1, 10:1.

Note A scale of 1 mm to 10 mm means th 1 mm on the drawing represents 10 mm on th object.

The scale to which the drawing has been draw should be stated as on Figures 7.10, 7.11 ar 7.12. Enlarging scales are used for drawir small objects larger than full size for clearnes

Dimensioning of scale drawings

The actual dimensions to which the object is t be made are shown on the drawings as in Fi ures 7.10, 7.11 and 7.12 irrespective of the sca used.

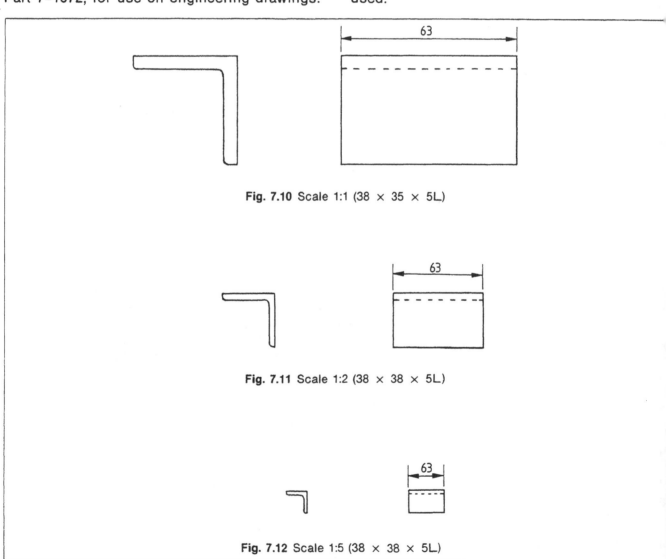

Fig. 7.10 Scale 1:1 (38 × 35 × 5L)

Fig. 7.11 Scale 1:2 (38 × 38 × 5L)

Fig. 7.12 Scale 1:5 (38 × 38 × 5L)

Metric scales

Note The distance *63 mm* reduces as the scale ratio increases.

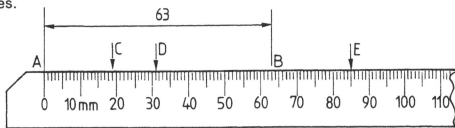

Fig. 7.13 Scale 1:1 (full size)

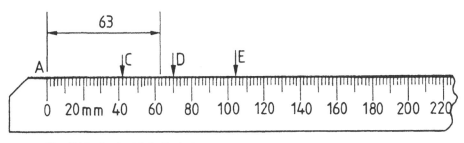

Fig. 7.14 Scale 1:2 (half size)

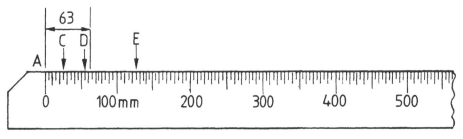

Fig. 7.15 Scale 1:5 (one-fifth size)

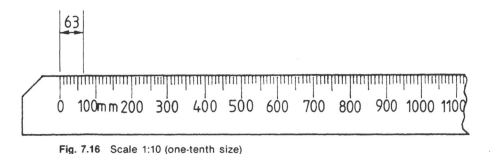

Fig. 7.16 Scale 1:10 (one-tenth size)

Exercise 1
Referring to Figure 7.13, what is the distance between points
A and C ?........ A and D ?........ A and E ? ....

Exercise 2
Referring to Figure 7.14, what is the distance between points
A and C ?........ A and D ?........ A and E ? ....

Exercise 3
Referring to Figure 7.15, what is the distance between points
A and C ?........ A and D ?........ A and E ?........

Exercise 4
Referring to Figure 7.16, mark on the scale rule the following distances: A 250, B 375, C 1005.

NOTES

8

Marking of cambers

The accurate marking of cambers is essential for the boilermaker marker-off.

Cambers are widely used in structural steelwork (supporting spans), shipbuilding and the development of large slow tapering conical work. In this chapter we will give you various methods presented in a step-by-step format to help you successfully lay out a true camber. Remember a true camber is a continuous curve and not a series of straight lines.

Note
Do not measure from the diagrams except in exercises, because many figures are indication diagrams only and are not drawn to scale. All measurements on the diagrams are in millimetres unless stated otherwise.

Cambers

Camber is the curving of a plate or section so that it is higher at the centre of its length than at the ends. The curvature should always represent a section of a true circle.

For a beam, the amount of camber is usually such that when the member is under maximum possible loading, it is straight.

Floor plate may be cambered to provide rigidity, reducing the number of supporting members required. The camber may also be needed to shed water, as in the deck plates of ships.

When forming the camber, its accuracy is checked by referring to a layout or a full or partial template. Large members may be checked by stretching a piano wire and measuring the amount of set at predetermined intervals. These degrees of set are called *ordinates*.

No matter which method is used, information is required which may be given on a drawing or which may need to be derived by the tradesman.

Given the span and degree of camber, the camber may be developed geometrically or by calculating the length of the ordinates.

Practical example

A cambered beam to allow for straightening when under load

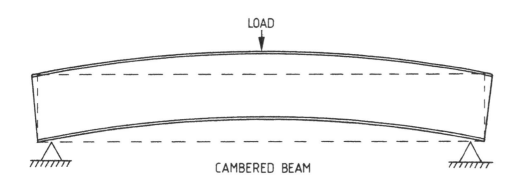

LOAD

CAMBERED BEAM

Two geometric methods for constructing cambered lines

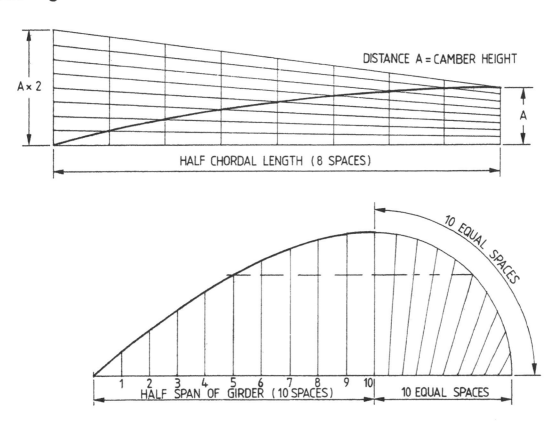

DISTANCE A = CAMBER HEIGHT

A × 2

A

HALF CHORDAL LENGTH (8 SPACES)

10 EQUAL SPACES

1 2 3 4 5 6 7 8 9 10
HALF SPAN OF GIRDER (10 SPACES)

10 EQUAL SPACES

Problem

To construct a camber using a very accurate method.

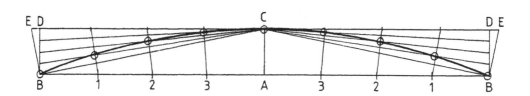

1	(a) Draw in base line *BB* equal in length to the camber. (b) Bisect the base line at point *A* and erect perpendicular line *AC* equal in height to the camber. (c) Join point *C* to *B*.

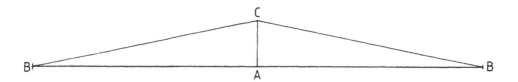

2	(a) Erect a line perpendicular to line *CB* from point *B*. (b) Draw in line parallel to *BAB* through point *C* to locate points *E*.

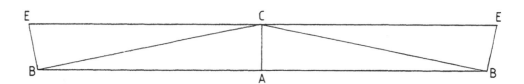

3	(a) Divide lines *BA* and *AB* into any number of equal spaces, e.g. say four. (b) Divide lines *EC* and *CE* into same amount of equal spaces. (c) Join all the division lines, 1, 2, 3.

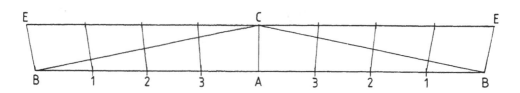

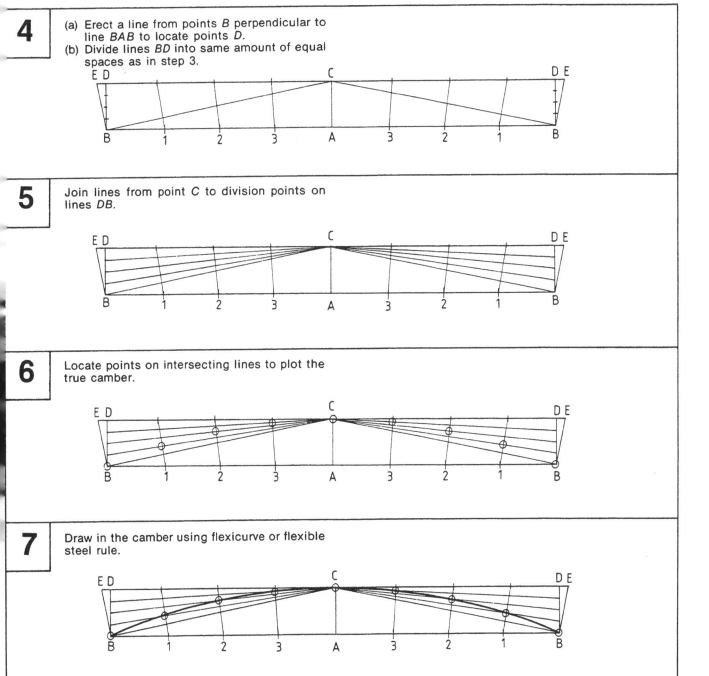

4

(a) Erect a line from points *B* perpendicular to line *BAB* to locate points *D*.
(b) Divide lines *BD* into same amount of equal spaces as in step 3.

5

Join lines from point *C* to division points on lines *DB*.

6

Locate points on intersecting lines to plot the true camber.

7

Draw in the camber using flexicurve or flexible steel rule.

Problem

To develop a camber using a very practical method.

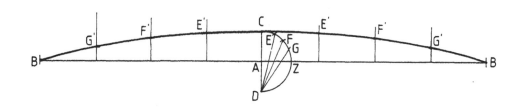

1	(a) Draw in base line *BB* equal to length of camber.
	(b) Bisect base line *BB* at point *A* and erect centre line *AC* perpendicular to base line *BB* equal to the height of the camber.

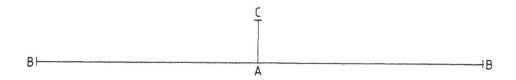

2	(a) Using point *A* as centre and *AC* as radius, scribe in half circle.
	(b) Extend centre line *AC* to locate point *D*.

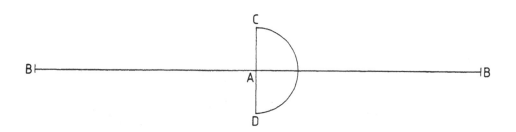

3	Divide the arc *CZ* into any number of equal spaces, say four, to locate the points *E*, *F*, *G*.

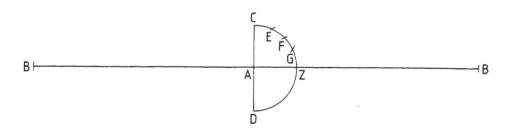

4 | Join point *D* to points *E, F, G* to locate points *H, I, J* on the base line.

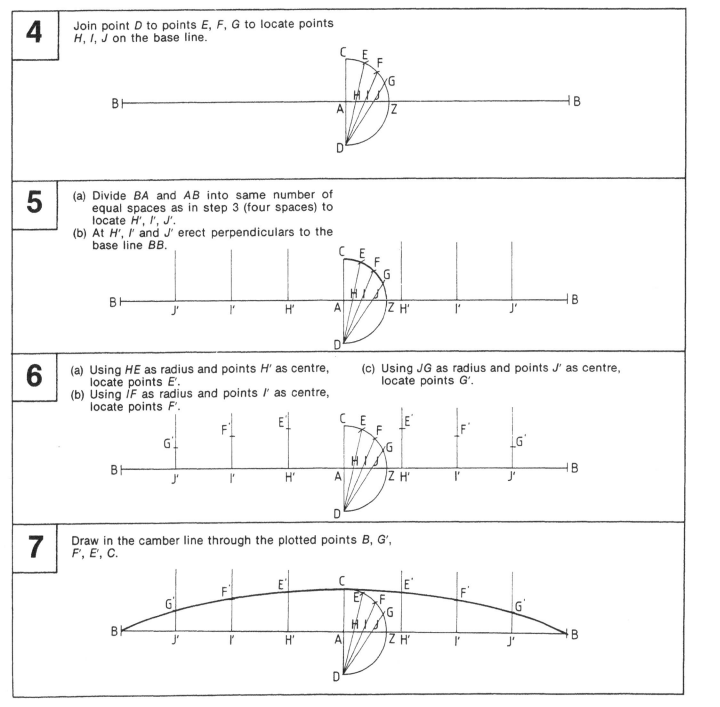

5 | (a) Divide *BA* and *AB* into same number of equal spaces as in step 3 (four spaces) to locate *H', I', J'*.
(b) At *H', I'* and *J'* erect perpendiculars to the base line *BB*.

6 | (a) Using *HE* as radius and points *H'* as centre, locate points *E'*.
(b) Using *IF* as radius and points *I'* as centre, locate points *F'*.
   (c) Using *JG* as radius and points *J'* as centre, locate points *G'*.

7 | Draw in the camber line through the plotted points *B, G', F', E', C*.

Problem

To construct a camber using another practical method.

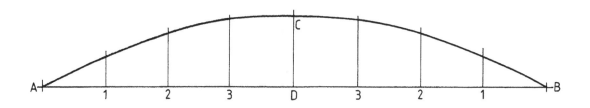

1

(a) Mark out length of camber to equal length of line *AB*.
(b) Bisect base line *AB* with centre line *CD*.
(c) Mark in height of camber (25 mm) on centre line *CD*.

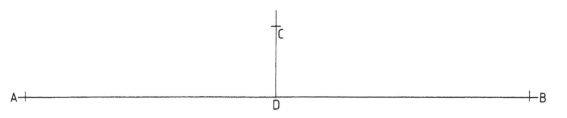

2

(a) Divide lines *AD* and *DB* into four equal sections and number the points 1, 2, 3.
(b) Erect vertical lines from points, 1, 2, 3.

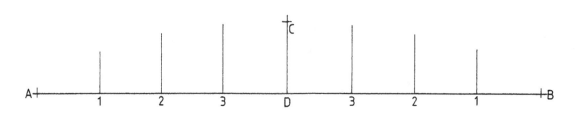

3

At points 1, mark in $^7/_{16}$ of camber height.
e.g. 25 mm camber $\therefore$ $^{25}/_{16} \times 7 = 10.9375$ mm.

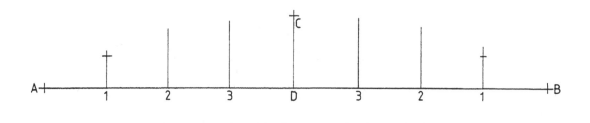

4

At points 2, mark in $^{12}/_{16}$ of camber height.
e.g. 25 mm camber $\therefore$ $^{25}/_{16} \times 12 = 18.75$ mm.

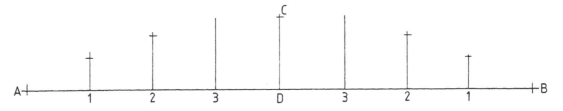

5

At points 3, mark in $^{15}/_{16}$ of camber height.
e.g. 25 mm camber $\therefore$ $^{25}/_{16} \times 15 = 23.4$ mm.

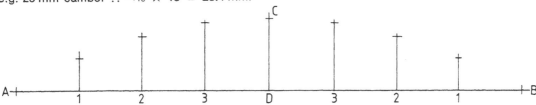

6

Mark in cambered line using flexicurve or flexible straight edge.

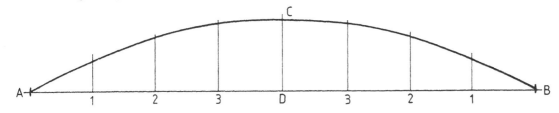

Explanation of extra lines for longer camber

Let line AB = height of camber
Let line BC = half length of camber
Divide BC into any number of equal spaces (say eight).

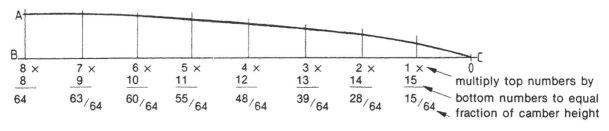

$8 \times$	$7 \times$	$6 \times$	$5 \times$	$4 \times$	$3 \times$	$2 \times$	$1 \times$	0	multiply top numbers by
8	9	10	11	12	13	14	15		bottom numbers to equal
64	$^{63}/_{64}$	$^{60}/_{64}$	$^{55}/_{64}$	$^{48}/_{64}$	$^{39}/_{64}$	$^{28}/_{64}$	$^{15}/_{64}$		fraction of camber height

| 15 mm | 14.7 mm | 14 mm | 12.89 mm | 11.25 mm | 9.14 mm | 6.56 mm | 3.51 mm | ordinate heights to be measured |

Analytical methods of determining curve of camber

Further explanation of previous camber method

1. Draw AB = half span of camber.
2. Divide AB into any number of equal spaces.
3. Number points in line ① from 1 to 4 starting with second point as shown. On line ② reverse the direction of numbering as shown.
4. Line ③ is the product of lines ① and ②.
5. The height of ordinates is the amount of the camber at centre line multiplied by a fraction whose numerator is the number on line ③ at the given point and whose denominator is the figure on line ③ at centre line, therefore:
   at point 1, camber is $\frac{7}{16} \times 40 = 17.5$
   at point 2, camber is $\frac{12}{16} \times 40 = 30$
   at point 3, camber is $\frac{15}{16} \times 40 = 37.5$

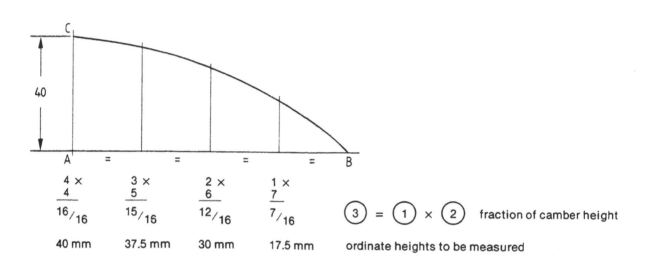

Practical examples

Sketch of cambered upper deck as used in shipbuilding

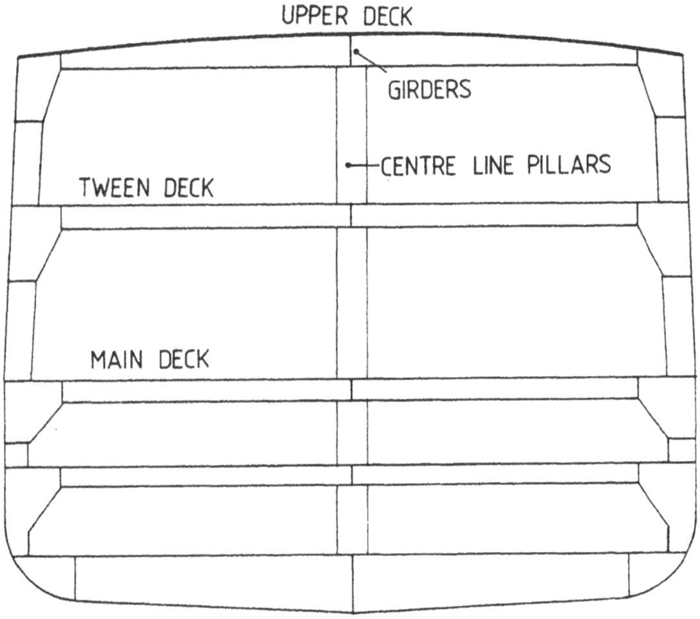

UPPER DECK

GIRDERS

CENTRE LINE PILLARS

TWEEN DECK

MAIN DECK

Drawing taken from structural drawing showing cambered floor beam

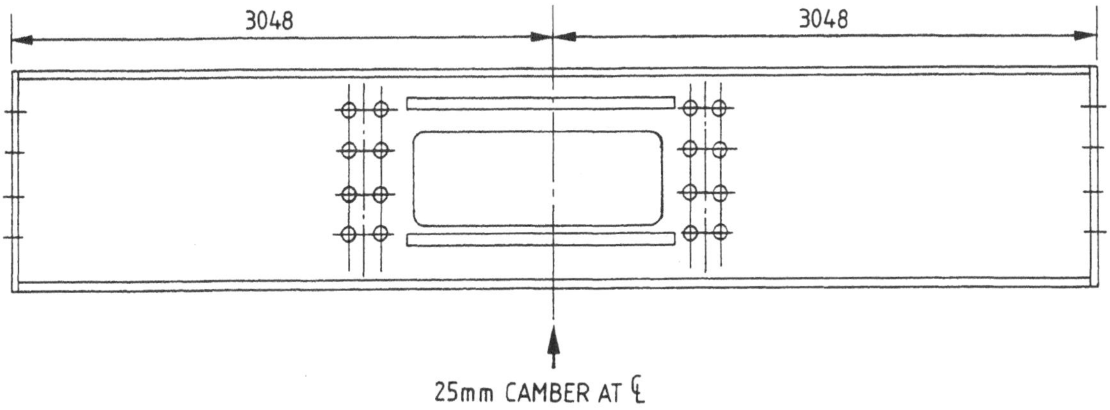

3048   3048

25mm CAMBER AT ₵

Practical method of determining curve of camber

1. Draw AB = full span of camber.
2. Construct C at mid-point height of camber.
3. Using timber or light section steel, make a frame which when braced and held rigid can be used to trace curve of camber by ensuring that the edges slide adjacent to points A and B.

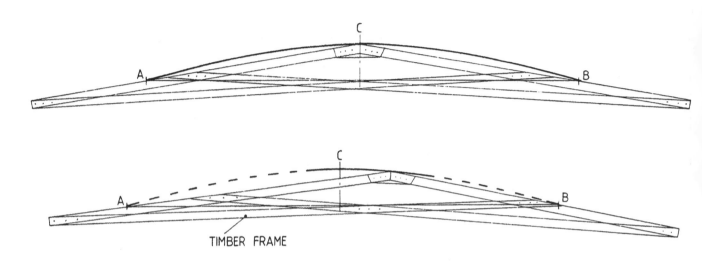

TIMBER FRAME

Practical method of determining curve of camber

1. Draw AB = full span of camber.
2. Construct CC_1 90° to AB on centre. $CC_1 = 2 \times$ camber height.
3. Divide AC and CB into the same number of equal spaces and number points as shown.
4. Join points 1, 1_1; 2, 2_1 etc., with straight lines.
5. Curve of camber is formed by drawing curved line to form a tangent at points shown. The more lines drawn, the easier it is to draw curve.

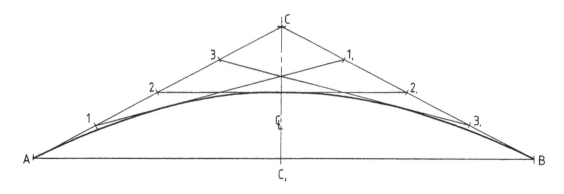

Practical use of a cambered line to develop pattern for slow tapering cone

Development of half camber line for half base curve of cone pattern

1. Draw *AB* = half chordal length.
2. Construct perpendicular *AC* = verse sine.
3. Construct *BD* perpendicular to *BA*.
4. Join diagonal *CB* and construct *BE* at 90° to *BC*.
5. Divide *AB*, *CE* and *BD* each into a number of equal parts, e.g. four.
6. Join points on *AB* to points on *CE*.
7. Join points on *BD* to *C*.
8. Join the points where these lines intersect with a smooth curve to form half base curve of developed cone.

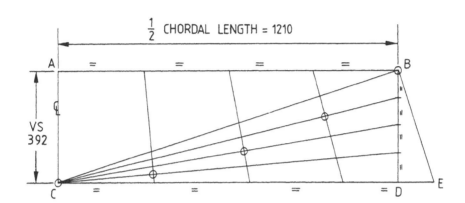

Development of cone pattern from constructed camber line

1. Develop half base curve as shown above.
2. Transfer to other half of pattern.
3. From centre line measure around half mean circumference of base.
4. From base curve, using sh frustum, draw a series of arcs up to plot top curve of frustum.
5. Draw a continuous curve tangent to these arcs, to form top curve of frustum.
6. From centre line, measure around half mean circumference of top.
7. Join extremities of top and bottom circumferences to form ends of cone.
8. Check diagonals.

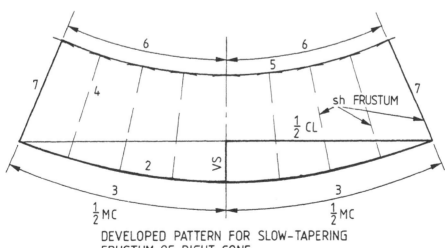

DEVELOPED PATTERN FOR SLOW-TAPERING
FRUSTUM OF RIGHT CONE

Exercises for constructing cambers

1. Construct a true camber using a very accurate method.

2. Construct a true camber using a very practical method.

3. Construct a true camber using another practical method.

9

Marking off an ellipse

Elliptical shapes are used widely for manhole and inspection openings in pressure vessels, road and rail tankers, and some tank work.

The boilermaker marker-off must be competent in the marking off of elliptical shapes for the correct fabrication of the above products.

This chapter shows the marking-off techniques for both true and approximate elliptical shapes.

The approximate elliptical shape is very practical as it allows the use of radial line marking using dividers and/or trammels. The centre points will allow cutting with a radius arm attachment and locate definite tangent points where change of radius takes place to assist the forming process.

Make sure the approximate method is acceptable for the job you are working on before you use this method.

Note
Do not measure from the diagrams except in exercises, because many figures are indication diagrams only and are not drawn to scale. All measurements on the diagrams are in millimetres unless stated otherwise.

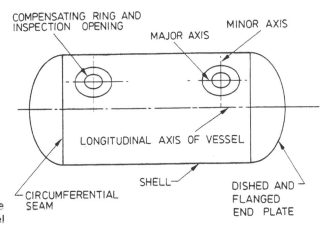

Fig. 9.1 Elliptical shaped inspection holes in a pressure vessel

Problem

To construct a true ellipse using the trammel method, when given the lengths of the major and minor axes.

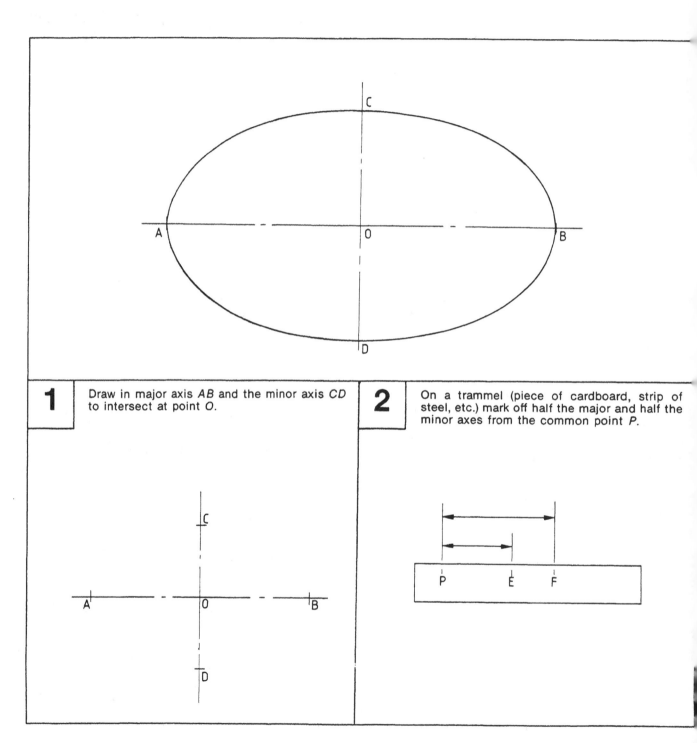

1 Draw in major axis *AB* and the minor axis *CD* to intersect at point *O*.

2 On a trammel (piece of cardboard, strip of steel, etc.) mark off half the major and half the minor axes from the common point *P*.

3 Using the trammel and lining up points *E* and *F* on the major and minor axes, mark in points at point *P* between points *A* and *C*.

4 Repeat the same procedure as in step 3 to locate the points between points *C* and *B*.

5 Repeat the same procedure on the lower half as in steps 3 and 4 to locate the points *A* to *B* through point *D*.

6 Join all the located points to form a continuous curved line which constructs a true ellipse.

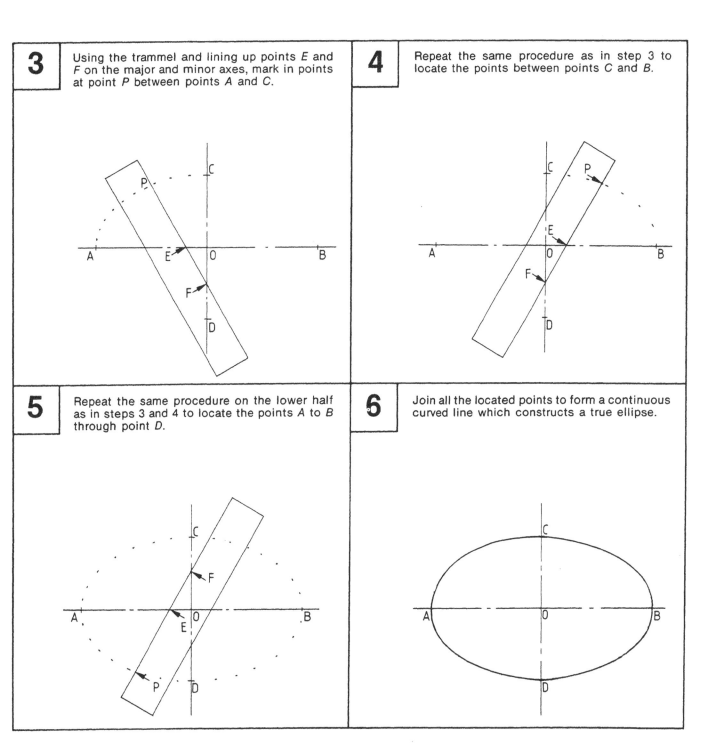

Problem

To construct an ellipse using concentric circles, given the minor and major axes.

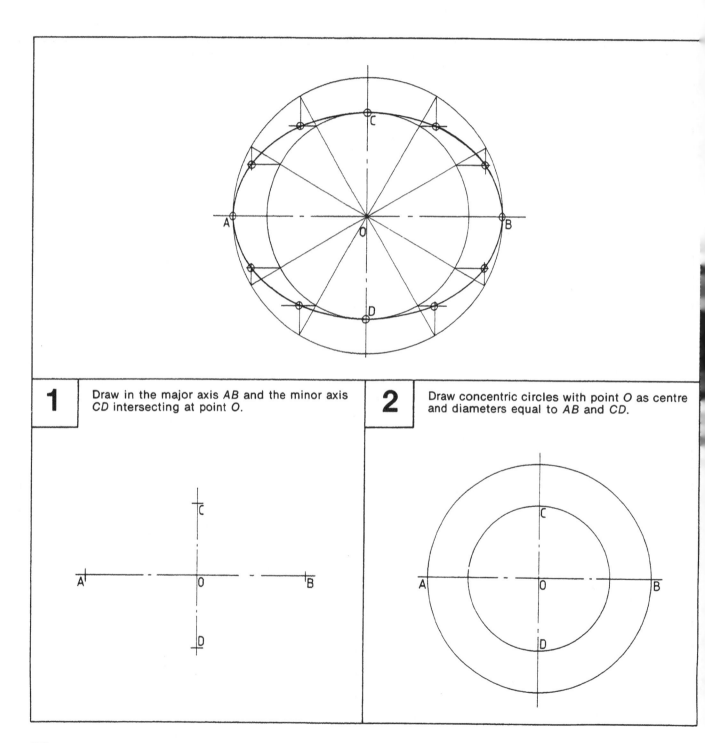

1	Draw in the major axis *AB* and the minor axis *CD* intersecting at point *O*.	**2**	Draw concentric circles with point *O* as centre and diameters equal to *AB* and *CD*.

3	Divide the large circle into twelve equal parts and draw in the diameters.	**4**	Draw lines parallel to the minor axis *CD* from the points where the diameters intersect the larger circle.

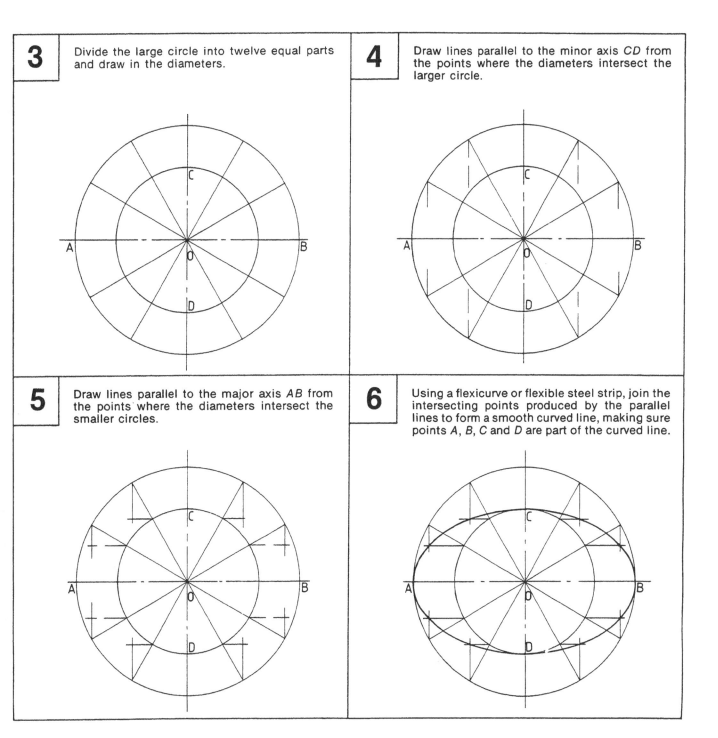

5	Draw lines parallel to the major axis *AB* from the points where the diameters intersect the smaller circles.	**6**	Using a flexicurve or flexible steel strip, join the intersecting points produced by the parallel lines to form a smooth curved line, making sure points *A*, *B*, *C* and *D* are part of the curved line.

Problem
To construct an ellipse using a rectangle.

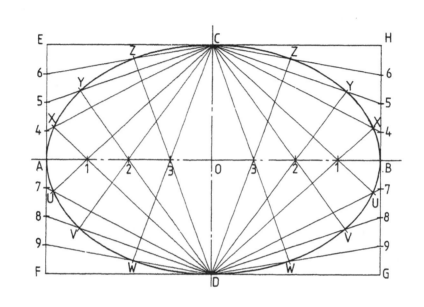

1	(a) Construct major and minor axes, *AB* and *CD*, to bisect at point O. (b) Construct rectangle *EFGH* around the minor and major axes' measurements.

2	(a) Divide *AO* into four equal parts to locate points 1, 2 and 3. (b) Divide *AE* into four equal parts to locate points 4, 5 and 6.

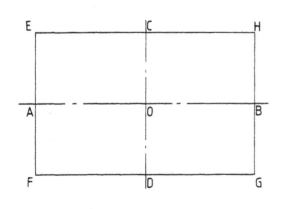

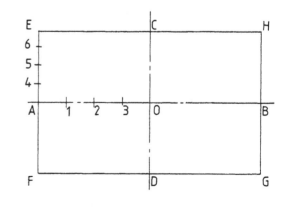

3	(a) Join *C* to 4, *C* to 5 and *C* to 6. (b) Join *D* to 1 and extend to locate point *X* on line *C4*. (c) Join *D* to 2 and extend to locate point *Y* on line *C5*. (d) Join *D* to 3 and extend to locate point *Z* on line *C6*.
4	(a) Divide *AF* into four equal parts to locate points 7, 8 and 9. (b) Join *D* to 7, *D* to 8 and *D* to 9. (c) Join *C* to 1, *C* to 2 and *C* to 3 and extend to locate points *U*, *V* and *W*.
5	Repeat steps 3 and 4 to locate the points *Z*, *Y* and *X* in rectangle *CHBO* and points *U*, *V* and *W* in rectangle *OBGD*.
6	Join all located points — *C*, *Z*, *Y*, *X*, *A*, *U*, *V*, *W*, *D* — with flexicurve or flexible straight edge to form the ellipse.

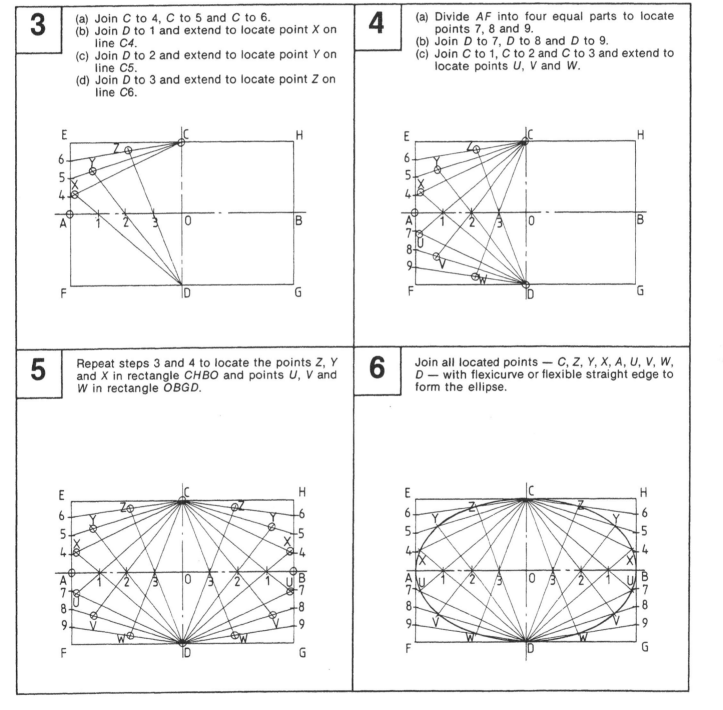

153

Problem

To construct an ellipse given the major and minor axes and using focal points, string and a marker.

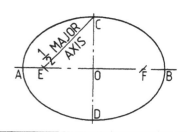

1	Draw in the major axis *AB* and the minor axis *CD* to intersect at point *O*.

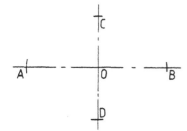

2	Using point *C* as centre and distance *AO* as radius, locate the focal points *E* and *F* on the major axis *AB*.

3	Fix pins to focal points *E* and *F* and attach string equal in length to major axis *AB* to these points. Keeping string taut, use marker to scribe a line from point *A* through point *D* to point *B*.

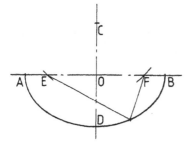

4	Using the same procedure as in step 3, scribe a curved line from point *A* through point *C* to point *B*. This then completes the construction of the ellipse.

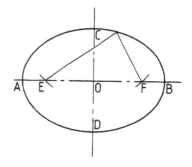

Practical example

Details of an elliptical inspection hole, compensating plate and cover plate as given on a workshop drawing.

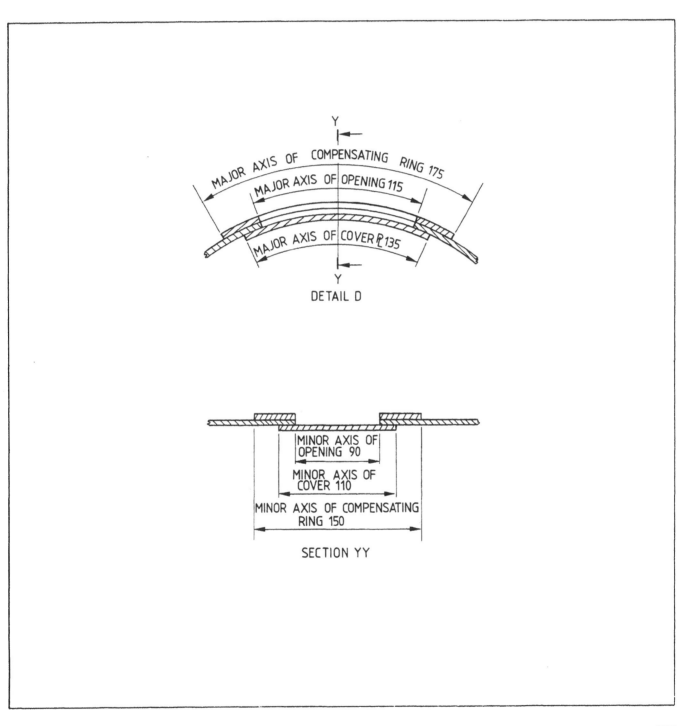

MAJOR AXIS OF COMPENSATING RING 175

MAJOR AXIS OF OPENING 115

MAJOR AXIS OF COVER ℓ 135

Y

Y

DETAIL D

MINOR AXIS OF OPENING 90

MINOR AXIS OF COVER 110

MINOR AXIS OF COMPENSATING RING 150

SECTION YY

Problem

To construct an approximate ellipse using four centres and radial lines.

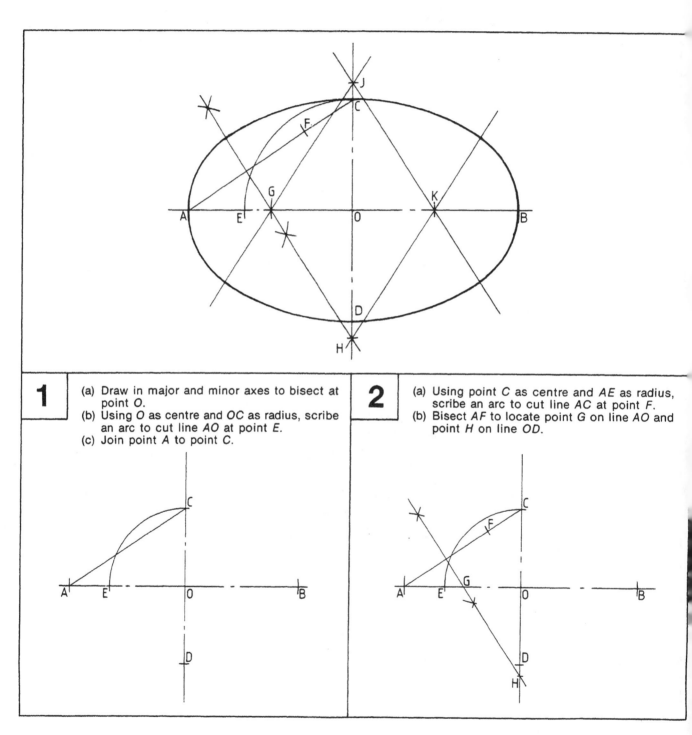

1
(a) Draw in major and minor axes to bisect at point O.
(b) Using O as centre and OC as radius, scribe an arc to cut line AO at point E.
(c) Join point A to point C.

2
(a) Using point C as centre and AE as radius, scribe an arc to cut line AC at point F.
(b) Bisect AF to locate point G on line AO and point H on line OD.

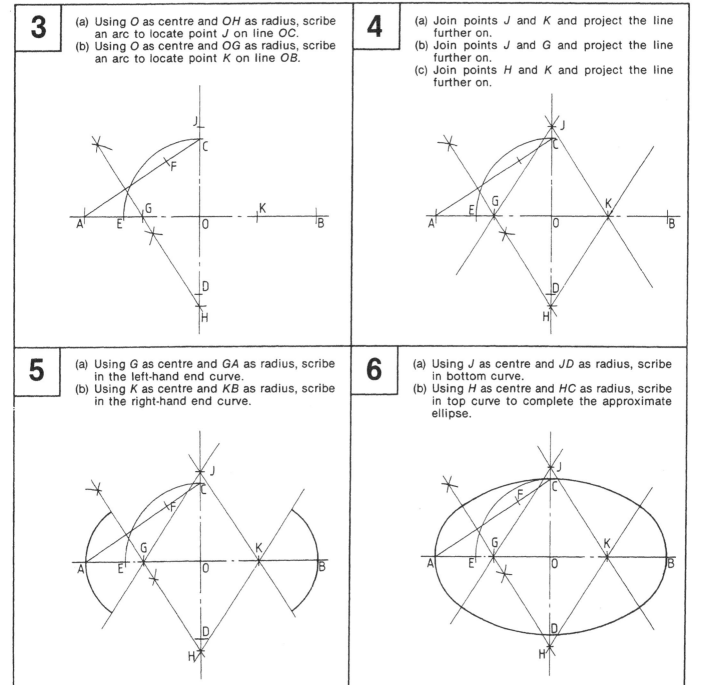

3
(a) Using *O* as centre and *OH* as radius, scribe an arc to locate point *J* on line *OC*.
(b) Using *O* as centre and *OG* as radius, scribe an arc to locate point *K* on line *OB*.

4
(a) Join points *J* and *K* and project the line further on.
(b) Join points *J* and *G* and project the line further on.
(c) Join points *H* and *K* and project the line further on.

5
(a) Using *G* as centre and *GA* as radius, scribe in the left-hand end curve.
(b) Using *K* as centre and *KB* as radius, scribe in the right-hand end curve.

6
(a) Using *J* as centre and *JD* as radius, scribe in bottom curve.
(b) Using *H* as centre and *HC* as radius, scribe in top curve to complete the approximate ellipse.

Problem

To construct an approximate ellipse using four centres and radial lines (alternative method).

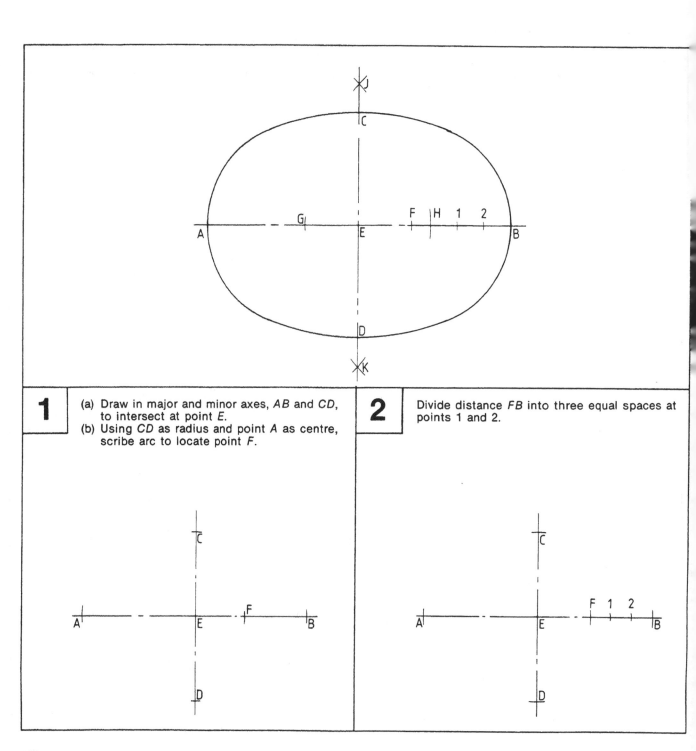

| **1** | (a) Draw in major and minor axes, *AB* and *CD*, to intersect at point *E*.
(b) Using *CD* as radius and point *A* as centre, scribe arc to locate point *F*. | **2** | Divide distance *FB* into three equal spaces at points 1 and 2. |

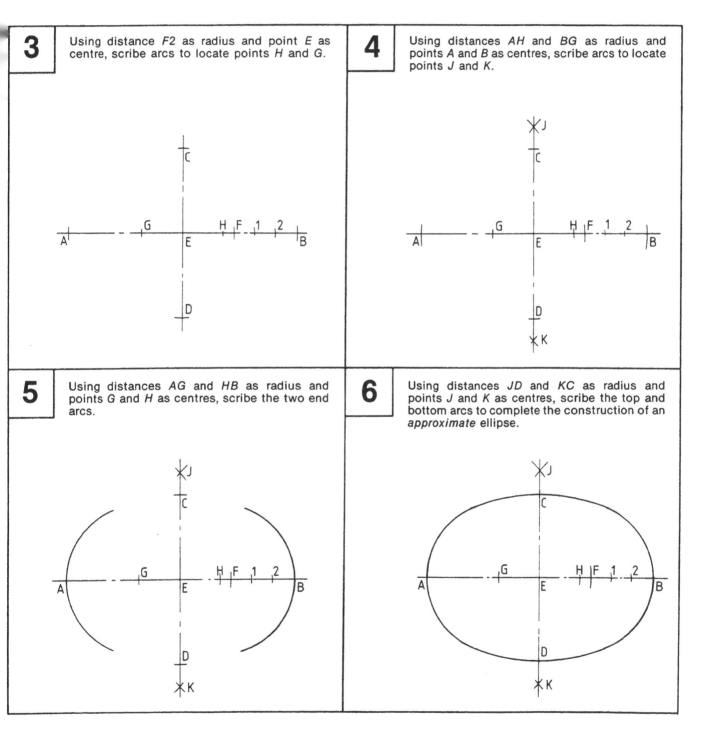

3 Using distance *F2* as radius and point *E* as centre, scribe arcs to locate points *H* and *G*.

4 Using distances *AH* and *BG* as radius and points *A* and *B* as centres, scribe arcs to locate points *J* and *K*.

5 Using distances *AG* and *HB* as radius and points *G* and *H* as centres, scribe the two end arcs.

6 Using distances *JD* and *KC* as radius and points *J* and *K* as centres, scribe the top and bottom arcs to complete the construction of an *approximate* ellipse.

Problem

To construct an approximate ellipse from a calculated radius of a circle (using four centre points to scribe the outline of the ellipse).

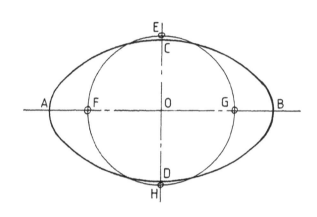

1

(a) Let centre line *AB* represent the major axis of the ellipse 62.5 mm in length.
(b) Let centre line *CD* represent the minor axis of the ellipse 38.5 mm in length.
(c) Let point *O* be the intersection point of the major and minor axes.

2

Calculate radius of circle.

radius = (major axis − minor axis) × 0.8536
= (62.5 − 38.5) × 0.8536
= 24 × 0.8536
= 20.4864 mm
say, 20.5 mm

3	Using point *O* as centre and calculated radius of 20.5 mm, scribe a circle to locate the four centre points *E*, *F*, *G*, *H*.	**4**	(a) Using point *F* as centre and distance *FA* as radius, scribe an arc. (b) Using point *G* as centre and distance *GB* as radius, scribe an arc.
5	Using point *H* as centre and distance *HC* as radius, scribe an arc tangent to the end arcs scribed in step 4.	**6**	Using point *E* as centre and distance *ED* as radius, scribe an arc tangent to the end arcs scribed in step 4 to complete the approximate ellipse.

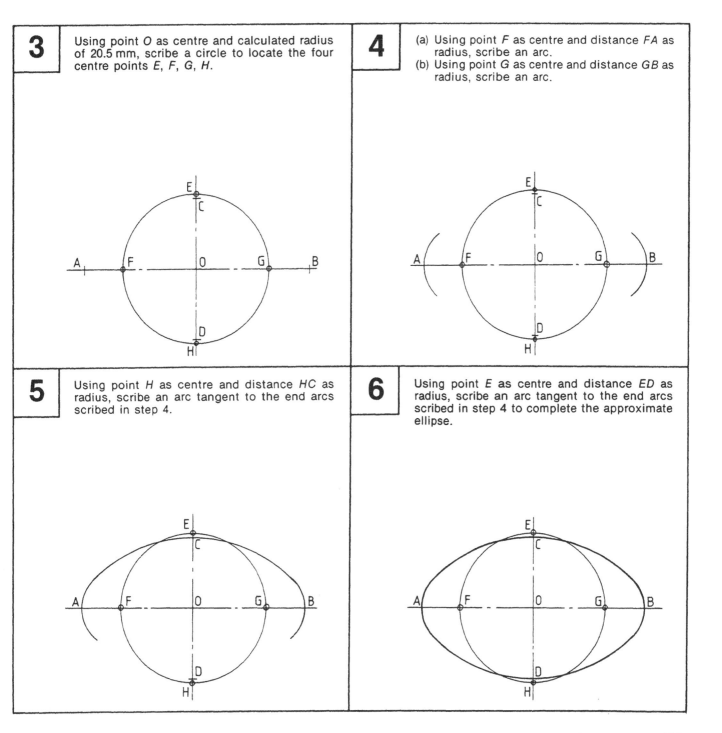

161

Exercises for constructing ellipses

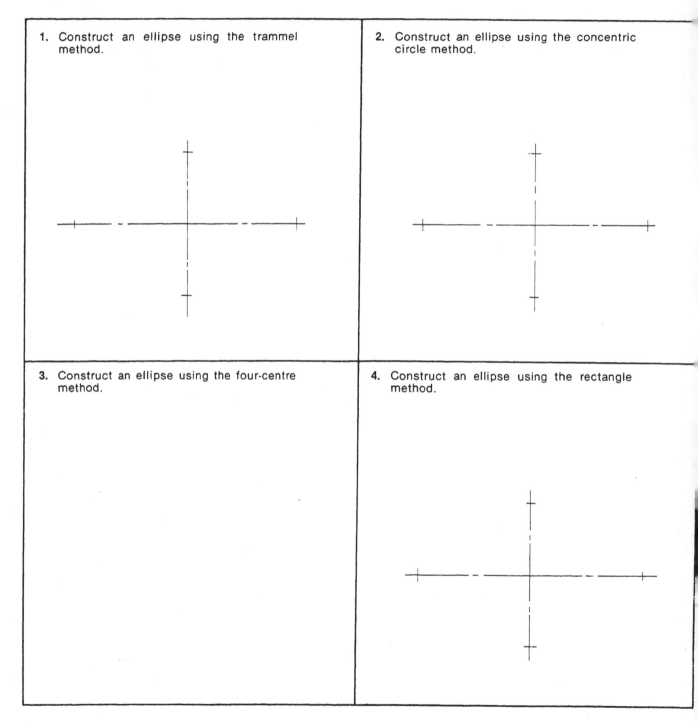

1. Construct an ellipse using the trammel method.

2. Construct an ellipse using the concentric circle method.

3. Construct an ellipse using the four-centre method.

4. Construct an ellipse using the rectangle method.

Fig. 9.2 Maxitherm D type water tube boiler showing use of elliptical shaped manholes in steam and mud drums

NOTES

10

Parallel line development

Parallel line development is used for the development of cylindrical shapes and pipework. To successfully mark out cylindrical and pipe work, a good understanding of inside diameter, outside diameter, mean diameter and mean circumference is required (refer to Chapter 6). In this chapter, we are endeavouring to cover parallel line development of intersecting cylinders, branch pipes and lobsterback pipe bends. These exercises will first be explained step by step, then progress to full development with an attached description page.

The following developments can be used when marking on flat plates before rolling, or when marking on formed pipes or cylinders with thin flat tinplate wrap around templates.

Thickness of material and types of joints will determine whether you will lay out to the inside, mean or outside diameters.

The following page shows the basic explanation of parallel line development, where the top and side views are divided into any number of equal spaces and the lengths from the side view are transferred to the corresponding lines of the development which is the circumference of the cylinder long and divided into the same number of equal spaces as the side and top views.

Note
Do not measure from the diagrams except in exercises, because many figures are indication diagrams only and are not drawn to scale. All measurements on the diagrams are in millimetres unless stated otherwise.

Development of a cylinder

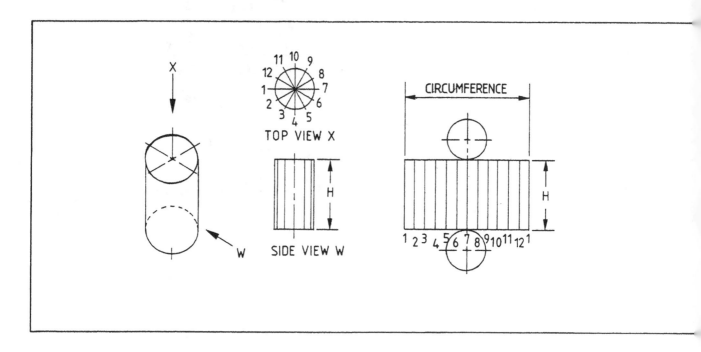

Development of a cylinder with one end cut at an angle

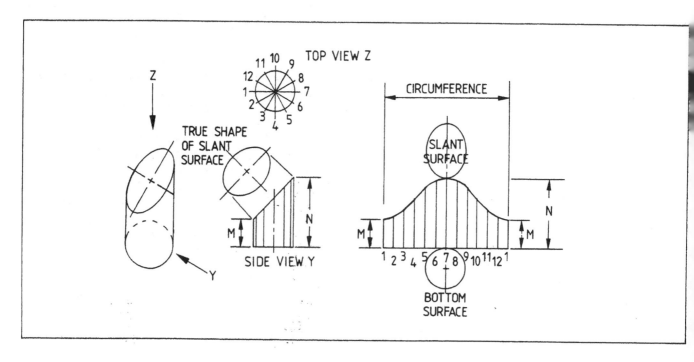

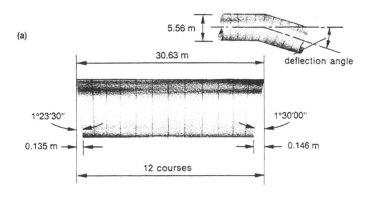

(a)

5.56 m

deflection angle

30.63 m

1°23′30″   1°30′00″

0.135 m   0.146 m

12 courses

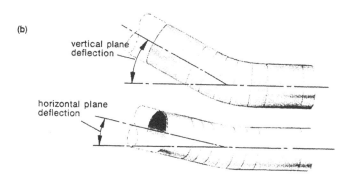

(b)

vertical plane
deflection

horizontal plane
deflection

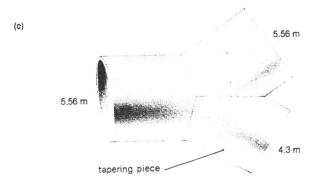

(c)

5.56 m

5.56 m

4.3 m

tapering piece

(d)

Fig. 10.1
Practical examples of pipework fabricated using
parallel line development: **(a)** Prefabricate
section of offset pipe bend **(b)** Double offset
elbow **(c)** Unequal bifurcate (Y piece) **(d)** 90°
lobsterback elbows

Problem

To develop a cylinder with one end cut at an angle.

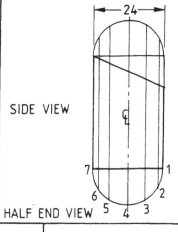

SIDE VIEW

HALF END VIEW

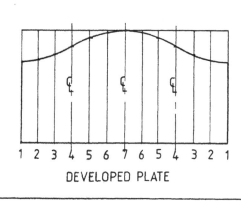

1 2 3 4 5 6 7 6 5 4 3 2 1

DEVELOPED PLATE

Calculation

$$MC = MD \times \pi$$
$$= 24 \times 3.1416$$
$$= 75.3984 \text{ mm}$$
$$\tfrac{1}{12} MC = 6.2832 \text{ mm}$$
$$\tfrac{1}{4} MC = 18.8496 \text{ mm}$$
$$\tfrac{1}{2} MC = 37.6992 \text{ mm}$$

1

(a) If marking off flat plate for cylinder, use mean diameter for the side view and mean circumference for the length of the developed plate prior to rolling.

(b) If marking off on to rolled or drawn pipe, use outside diameter for the side view and outside circumference for development lines and wrap-around templates.

(c) If marking off a pipe using a contour marker, use outside diameter for side view; the half end view or the parallel development lines are not needed.

2

(a) Draw side view to MD.

(b) Draw in half end view and divide into six equal spaces and number the points 1 to 7.

(c) Project points 2 to 6 from half end view to side view parallel to the centre line.

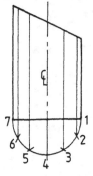

| **3** | (a) Construct a rectangle for the true developed shape, length to equal the mean circumference and the width to equal the length of line 7 from the side view. | (b) Divide the rectangle into twelve equal spaces and number the lines the same as the side view, using the end lines (seam) of the rectangle as line no. 1. |

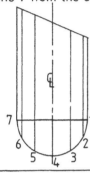 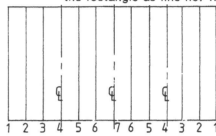

| **4** | Transfer the length of the lines from the side view on to the corresponding lines of the development. |

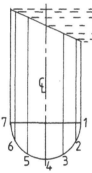 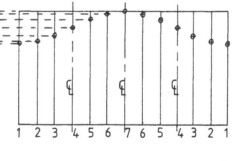

| **5** | Draw in the curved cut line with a flexicurve or flexible straight edge to complete the true development. |

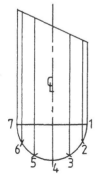 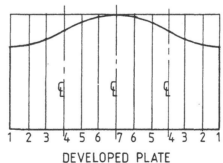

DEVELOPED PLATE

Vertical cylindrical branch on inclined flat plate
Operation sequence

Top and front view

1. Draw in front view.
2. Draw in top view and divide it into twelve equal spaces numbered 1 to 12.
3. Project points 2 to 12 parallel to centre line on to front view and label the points.

Developed pattern for branch pipe

1. Calculate the MC.

   Calculation

$$MC = MD \times \pi$$
$$= 70 \times 3.1416$$
$$= 219.912 \text{ mm}$$
$$\tfrac{1}{2} MC = 109.956 \text{ mm}$$
$$\tfrac{1}{4} MC = 54.978 \text{ mm}$$
$$\tfrac{1}{12} MC = 18.326 \text{ mm}$$

2. Mark out the development rectangle MC long by longest line of front view wide. (Check measurements and diagonals.)
3. Divide length of rectangle into twelve equal spaces and label them the same as in the front view.
4. Transfer lengths of lines from front view to corresponding lines of the development.
5. Draw in curved line using flexible straight edge to complete the developed shape of the vertical branch.

Developed pattern of hole in plate

1. Project square plate off inclined layout of front view and mark in centre lines *XY* and *VW*.
2. Project points 1 to 7 from inclined line, parallel to centre line *VW*.
3. Mark in lines equal to lines in top view, parallel to centre line *XY*, to form lines *K* to *D* on hole pattern in base plate.
4. Transfer points 1 to 7 from base line of front view and 90° to same inclined base line.
5. Intersecting lines from steps 11 and 12 will form the shape of hole in plate. (Mark in with flexible straight edge.)

Development of hole and pattern for vertical branch on inclined flat plate

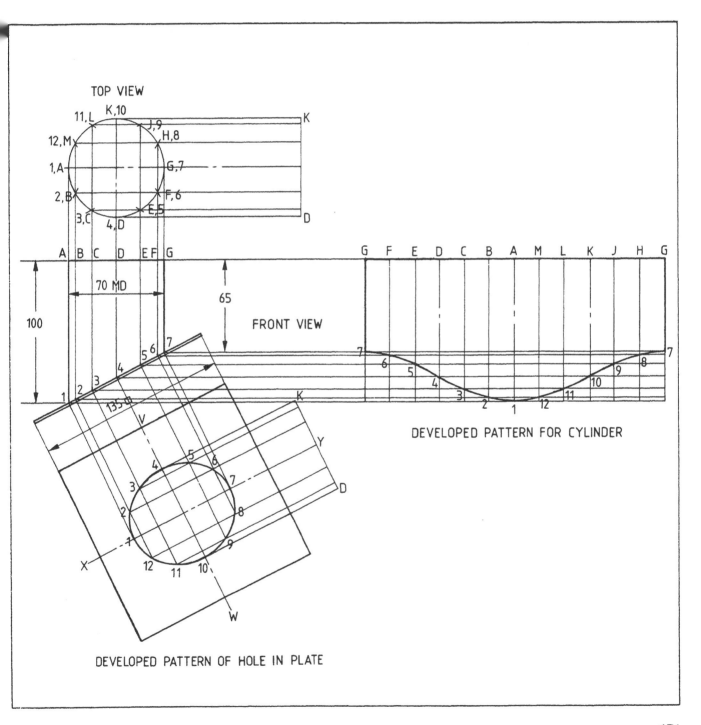

TOP VIEW

FRONT VIEW

DEVELOPED PATTERN FOR CYLINDER

DEVELOPED PATTERN OF HOLE IN PLATE

Equal diameter intersecting cylinders
Operation sequence

Front view
1. Draw in centre lines at correct angles and lengths.
2. Square off end lines.
3. Bisect the angle.
4. Measure in the correct diameter, distances and draw in the outlines parallel to the centre lines.
5. Draw in the half end views and divide them into, say, six equal spaces and number the points 1 to 7.
6. Draw in the lines 2 to 6 parallel to the centre lines.

Development of true shapes A *and* B

1. Calculate the MC of the cylinder.

   Calculation

   $$MC = MD \times \pi$$
   $$= 25 \times 3.1416$$
   $$= 78.54 \text{ mm}$$
   $$\tfrac{1}{2} MC = 39.27 \text{ mm}$$
   $$\tfrac{1}{4} MC = 19.635 \text{ mm}$$
   $$\tfrac{1}{12} MC = 6.545 \text{ mm}$$

2. Mark out the rectangles for the developments, equal to MC long and a height greater in length than longest point on each pipe *A* and *B* wide. (Check measurements and diagonals.)
3. Divide length of rectangles into twelve equal sections and number the lines the same as the side view.
4. Transfer the lengths of line 1 to 7 and *a* to *g* from the side view to the corresponding lines of the developments for cylinders *A* and *B*.
5. Draw in the curved lines with a flexible straight edge to complete the true developed shapes for cylinders *A* and *B*.

Note If size of plate and width of rolls are adequate, the cylinders *A* and *B* can be developed as a single plate and cut after rolling. (Similar to the next job offset pipe.)

Development of equal diameter intersecting cylinders

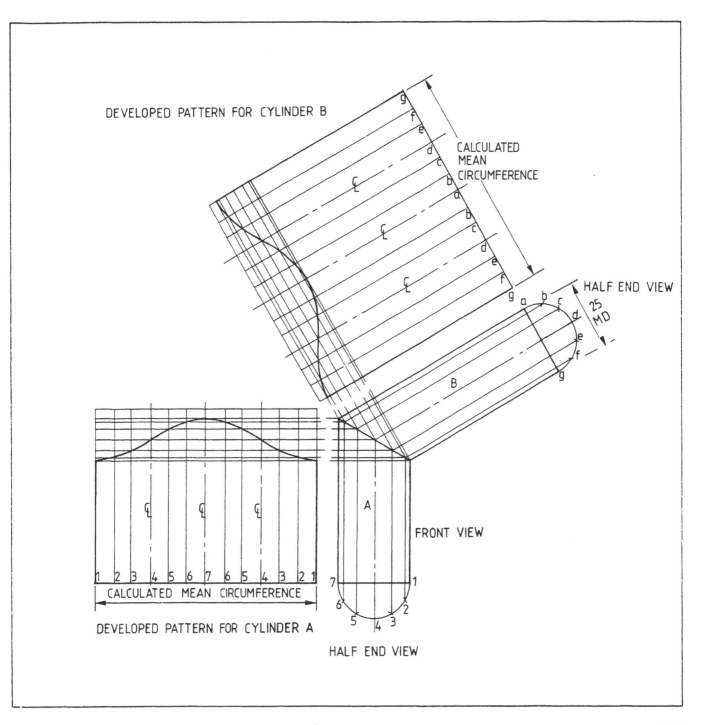

DEVELOPED PATTERN FOR CYLINDER B

CALCULATED
MEAN
CIRCUMFERENCE

HALF END VIEW

25
MD

B

A

FRONT VIEW

CALCULATED MEAN CIRCUMFERENCE

DEVELOPED PATTERN FOR CYLINDER A

HALF END VIEW

Offset pipe line

Operation sequence

1. Locate accurately the centre line of each cylinder.
2. At convenient distance away from centre lines, scribe semicircle with diameter square to centre line.
3. Draw parallel lines representing mean diameter of each cylinder. (Check correct position of centre lines and mean diameters.)
4. Draw mitre lines by bisecting the angles of the centre lines.
5. Add construction lines by dividing semicircles into six equal divisions and drawing parallel lines.
6. Calculate mean circumference of cylinder.
7. Commence development by constructing rectangle, mean circumference in length and sum of centre lines as height.
8. Divide length into twelve equal parts and number as shown.
9. Plot curves by transferring lengths from side view to corresponding lines on development.
10. Mark clearly centre lines and add other production instructions.
11. If capacity of rollers will permit, the three parts to be rolled as one cylinder and oxy cut after rolling.

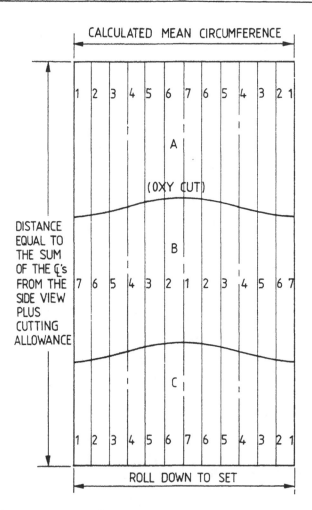

Development of cylinders *A*, *B* and *C* on one plate, if size of plate and width of rolls will allow this. (Oxy cut the 2 seams after rolling.)

Development as single cylinders

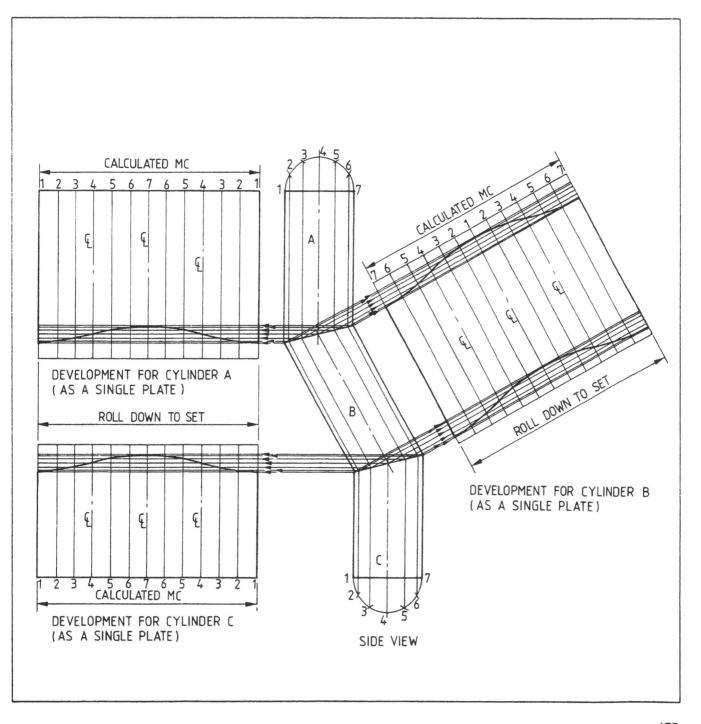

CALCULATED MC

1 2 3 4 5 6 7 6 5 4 3 2 1

DEVELOPMENT FOR CYLINDER A
(AS A SINGLE PLATE)

ROLL DOWN TO SET

DEVELOPMENT FOR CYLINDER C
(AS A SINGLE PLATE)

CALCULATED MC

1 2 3 4 5 6 7 6 5 4 3 2 1

A

B

C

SIDE VIEW

CALCULATED MC

7 6 5 4 3 2 1 2 3 4 5 6 7

DEVELOPMENT FOR CYLINDER B
(AS A SINGLE PLATE)

ROLL DOWN TO SET

Practical example

Job taken from a workshop drawing showing an
application of an offset pipe bend.

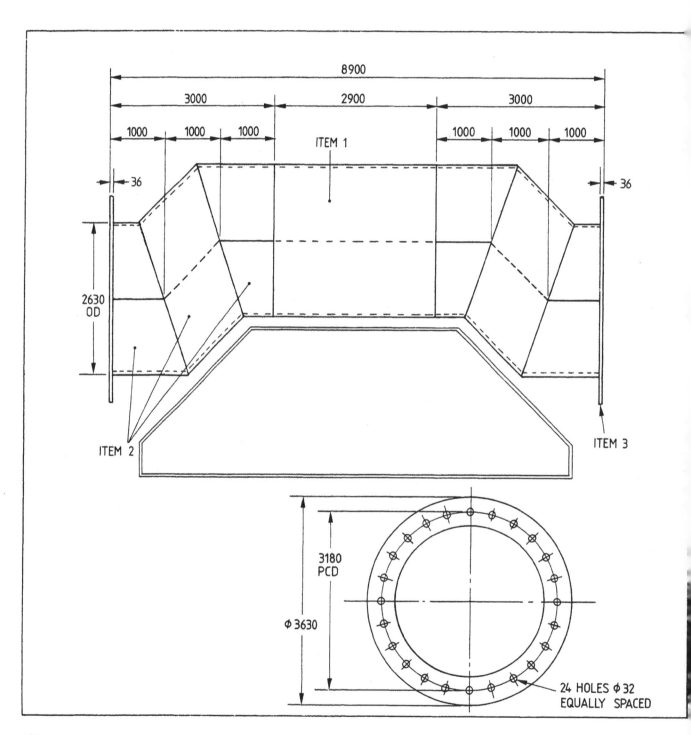

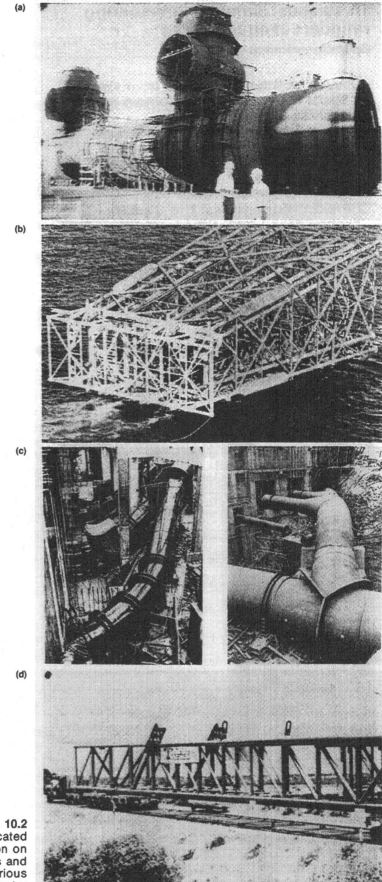

Fig. 10.2
Industrial applications of pipework fabricated using parallel line development: **(a)** T section on oil rig leg **(b)** Oil rig with cylindrical sections and various joints **(c)** Bifurcate (Y piece) **(d)** Various types of pipe joints on gantry

Three equal diameter intersecting cylinders at equal angles

Operation sequence

1. Draw in centre lines of the front view.
2. Bisect the three angles to obtain correct intersection lines.
3. Mark in the outlines of the cylinders.
4. Mark in the end lines of the cylinders.
5. Draw in the half end views on the end lines.
6. Divide the half end views into six equal spaces and number them 1 to 7.
7. Project division lines 2 to 6 parallel to the centre lines to meet the intersection lines.
8. Calculate the MC.

   Calculation

   $$MC = MD \times \pi$$
   $$= 400 \times 3.1416$$
   $$= 1256.64 \text{ mm}$$
   ½ MC = 628.32 mm
   ¼ MC = 314.16 mm
   ¹⁄₁₂ MC = 104.72 mm

9. Mark out the development rectangles, M long by the longest point from the front vie wide. (Check the measurements and th diagonals.)
10. Divide length of rectangles into twelve equa spaces and number them the same as th half end views from the front view.
11. Transfer lengths of lines from front view t corresponding line on the developments.
12. Draw in the curved lines using a flexib straight edge to complete the development for cylinders A, B and C.

Fig. 10.3 Breeching pieces with equal angles and equal diameters

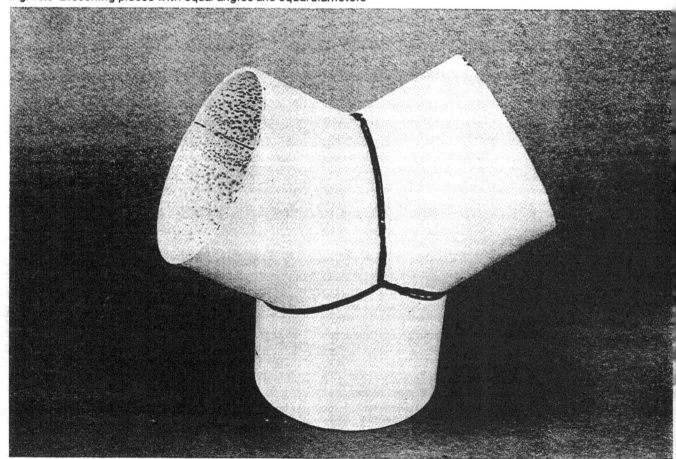

Development of three equal diameter intersecting cylinders at equal angles

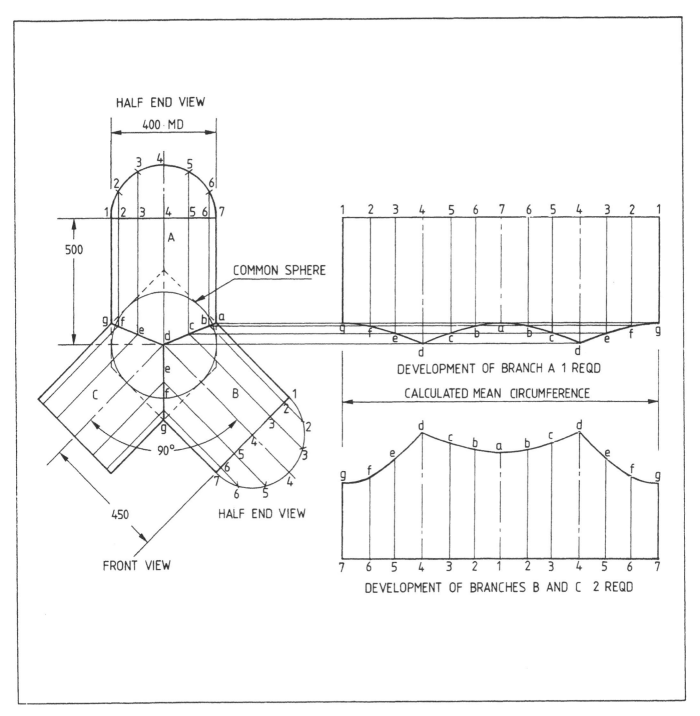

HALF END VIEW

400 · MD

COMMON SPHERE

A

500

DEVELOPMENT OF BRANCH A 1 REQD

CALCULATED MEAN CIRCUMFERENCE

C

B

90°

450

HALF END VIEW

FRONT VIEW

DEVELOPMENT OF BRANCHES B AND C 2 REQD

Three equal diameter intersecting cylinders with unequal angles

Operation sequence

1. Draw in centre lines of the front view.
2. Bisect the three angles to obtain the correct intersection lines.
3. Mark in outlines of the cylinders.
4. Mark in end lines of the cylinders.
5. Draw in the half end views of the cylinders on the end lines.
6. Divide the half end views into six equal spaces and number them 1 to 7.
7. Project the division lines 2 to 6 parallel to centre lines to meet intersection lines.
8. Calculate the MC.

   Calculation

   $$MC = MD \times \pi$$
   $$= 380 \times 3.1416$$
   $$= 1193.808 \text{ mm}$$
   $$\tfrac{1}{2} MC = 596.904 \text{ mm}$$
   $$\tfrac{1}{4} MC = 298.452 \text{ mm}$$
   $$\tfrac{1}{12} MC = 99.484 \text{ mm}$$

9. Mark out the development rectangles, M long by longest point from the front vie wide. (Check measurements and diagonals
10. Divide the length of the rectangles int twelve equal spaces and number them th same as the half end views from the fror view.
11. Transfer lengths of lines in front view to co responding lines on developments.
12. Draw in curved lines using flexible straigh edge to complete the developments.

Development of three equal diameter intersecting cylinders with unequal angles

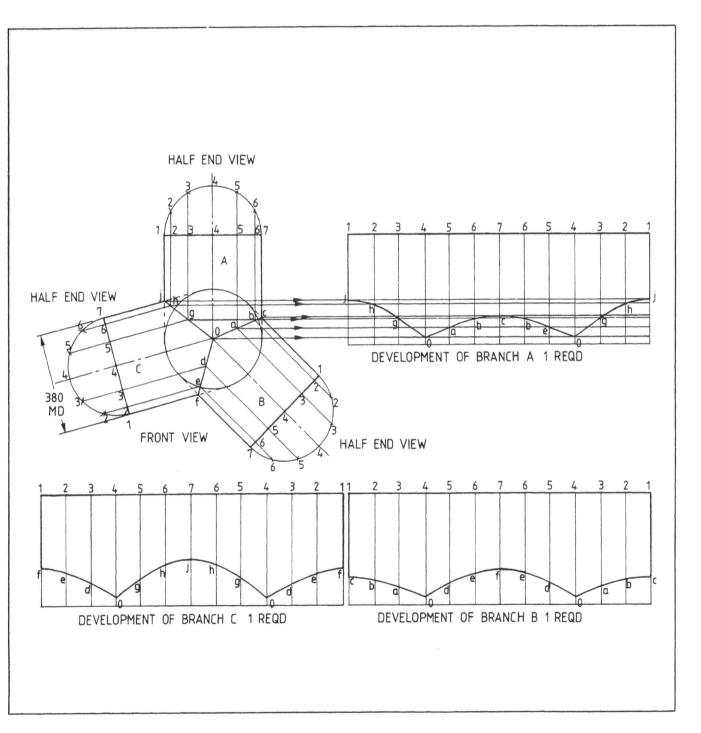

HALF END VIEW

HALF END VIEW

A

C

B

380 MD

FRONT VIEW

HALF END VIEW

DEVELOPMENT OF BRANCH A 1 REQD

DEVELOPMENT OF BRANCH C 1 REQD

DEVELOPMENT OF BRANCH B 1 REQD

Y piece — equal diameters
(centre section D)
Operation sequence

1. Draw in centre lines of the Y piece.
2. Bisect the three angles to obtain correct intersection lines.
3. Mark in the outlines of the cylinders.
4. Draw in front view of cylinder D.
5. Draw in half end view on end of cylinder D.
6. Divide half end view into six equal spaces and number them 1 to 7.
7. Project the division lines 2 to 6 parallel to the centre line.
8. Mark in reference line X on front view of cylinder D.
9. Calculate the MC.

   Calculation

   $$MC = MD \times \pi$$
   $$= 30 \times 3.1416$$
   $$= 94.248 \text{ mm}$$
   $$\tfrac{1}{2} MC = 47.124 \text{ mm}$$
   $$\tfrac{1}{4} MC = 23.562 \text{ mm}$$
   $$\tfrac{1}{12} MC = 7.854 \text{ mm}$$

10. Mark out the development rectangle, MC long by a width wider than the longest line from the front view D.
11. Mark in reference line X on development rectangle.
12. Divide length of rectangle into twelve equal spaces and number them the same as the half end view from the front view.
13. Transfer lengths of lines from front view to corresponding lines on development. (Measure each side of reference line X.)
14. Draw in curved lines using flexible straight edge to complete the development.

Development of centre section of
Y piece

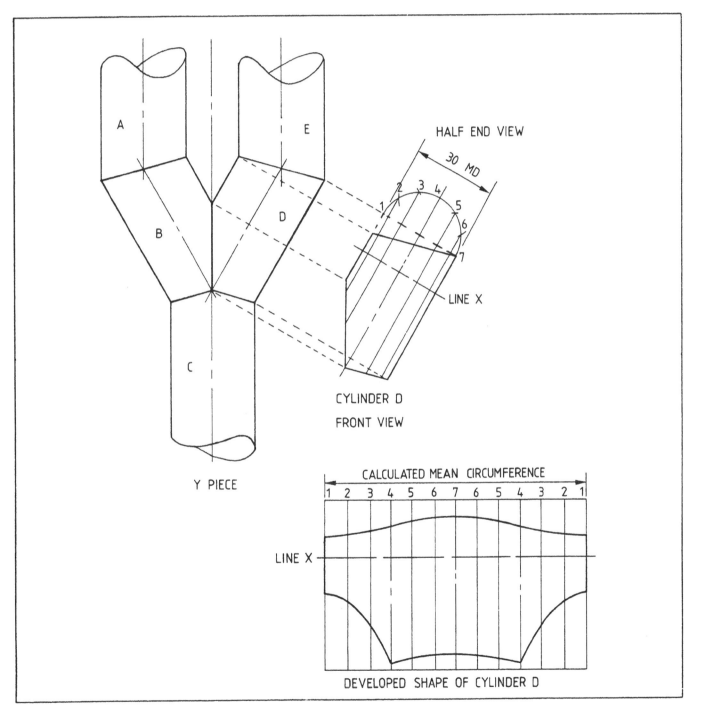

HALF END VIEW

30 MD

LINE X

CYLINDER D
FRONT VIEW

Y PIECE

CALCULATED MEAN CIRCUMFERENCE

LINE X

DEVELOPED SHAPE OF CYLINDER D

Considerations you must apply when laying out branch pipes:

1. Difference in diameters of the connections and the type of joint, as in Details *A*, *B* and *C*.
2. Thickness and position of branch wall, as in Details *D*, *E*, *F* and *G*.
3. Flat plate development.
4. Wrap around templates.
5. Marking off on outside of formed pipes and cylinders.

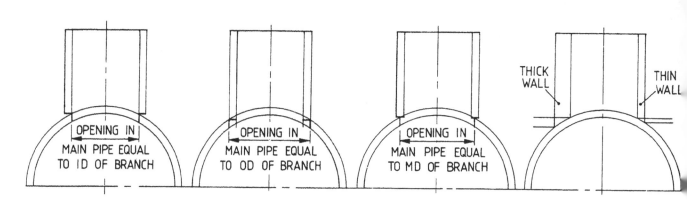

DETAIL A

Use ID for branch pipe and OD for main pipe.

DETAIL B

Use OD for branch pipe and main pipe.

DETAIL C

Use MD for branch pipe and OD for main pipe.

DETAIL D

Notes
1. Irrespective of what diameter you use for laying out, the same principle is applied for single line drawings (no thickness shown) of the side views and the development.
2. For flat plate development, use the mean circumference for lengths of plates for branches, etc.
3. For wrap around templates, use [(outside circumference of pipe) + (thickness of template material × π)] for length of wrap around template.

These three details, *E*, *F* and *G*, show the vast differences that can occur in branch connections between inside and outside penetration points of the branch to the main pipe.

These differences coupled with different thickness branches and types of joints in Details *A*, *B*, *C* and *D* on the previous page, show why it is necessary to understand what diameters to use when laying out branch connections.

The rest of the layouts and developments in this chapter will feature all single line drawings and you will have to apply these to your particular needs for each individual connection.

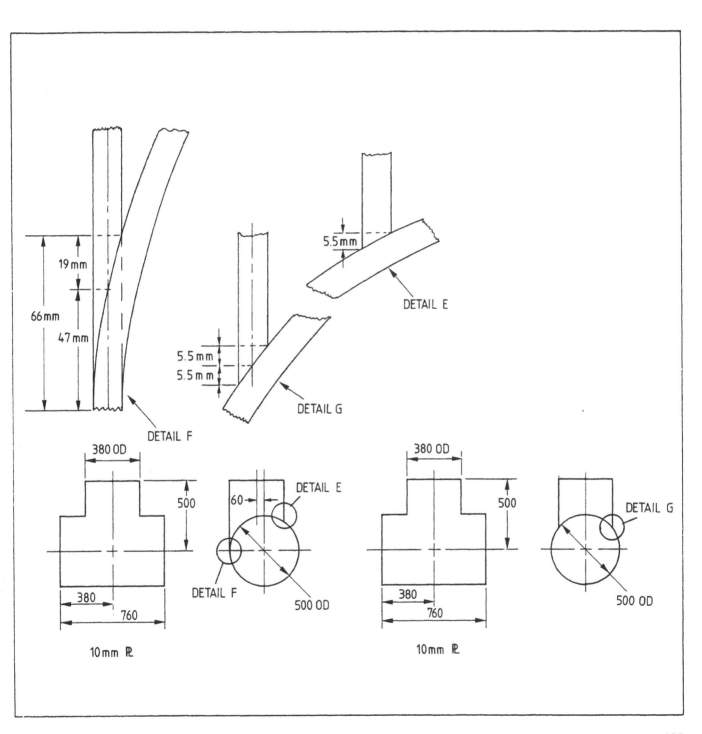

185

Cylindrical branch on centre and 90° to the horizontal centre line (unequal diameters)

Operation sequence

Front and end views

1. (a) Draw in left end view to mean dimensions.
   (b) Draw in development lines from half top view and number 1 to 7.
2. (a) Draw in front view projected from left end view.
   (b) Draw in development lines from half top view and number them accordingly.
3. Draw in line of penetration.
   (a) Project points from end view to corresponding numbered lines in front view.
   (b) Draw in line of penetration through the plotted points.

Development of cylinder B

1. Calculate the MC of the branch.

   Calculation

   $$\text{MC of } B = \text{MD} \times \pi$$
   $$= 20 \times 3.1416$$
   $$= 62.832 \text{ mm}$$

½ MC	= 31.416 mm
¼ MC	= 15.708 mm
¹⁄₁₂ MC	= 5.236 mm

2. Construct rectangle by projecting from front view, mean circumference long by distance longer than lines 7 and 1 wide.
3. Divide length of rectangle (MC) into twelve equal spaces.
4. Number the division lines using the short line 4 as the seam.
5. Project lengths of lines for branch (to line of penetration) on to the corresponding lines of the branch development.
6. Join transferred points with curved lines to complete the development of the branch.

Development of cylinder A

1. Calculate the MC of the main cylinder.

   Calculation

   $$\text{MC of } A = \text{MD} \times \pi$$
   $$= 30 \times 3.1416$$
   $$= 94.248 \text{ mm}$$

½ MC	= 47.124 mm
¼ MC	= 23.562 mm
¹⁄₁₂ MC	= 7.854 mm

2. Construct rectangle by projecting from fro view, MC long by length of front view wid
3. Divide length of rectangle into four equ spaces to locate the centre lines.
4. Project the development lines from the ha top view of the front view, across centre lir 4 of the rectangle (cylinder A).
5. Transfer the curved distances from the le end view points a, b, c and d and locate the either side of centre line 4 of the develo ment.
6. Where the points a, b, c and d cross the pr jected lines from the front view, they loca the points for the true developed shape the hole.
7. Join these points with a curved line complete the true developed shape of th opening.
8. Mark the branch pipe B and the main pipe Roll down to correct inside diameter.

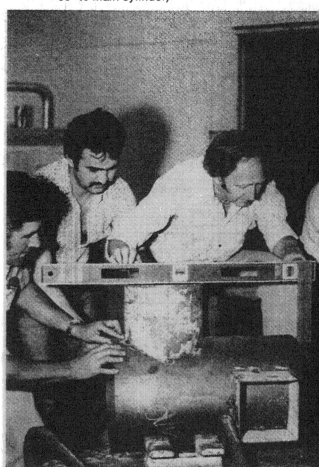

Fig. 10.4 Checking the job (round branch on centre and a 90° to main cylinder)

Development of cylindrical branch on centre and 90° to the horizontal centre line (unequal diameters)

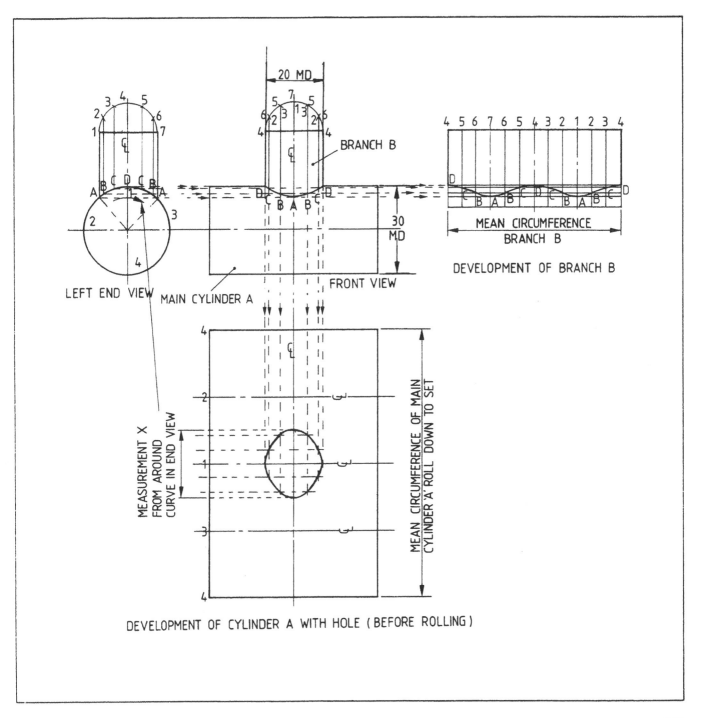

LEFT END VIEW

MAIN CYLINDER A

20 MD

BRANCH B

30 MD

FRONT VIEW

4 5 6 7 6 5 4 3 2 1 2 3 4

MEAN CIRCUMFERENCE
BRANCH B

DEVELOPMENT OF BRANCH B

MEASUREMENT X FROM AROUND CURVE IN END VIEW

MEAN CIRCUMFERENCE OF MAIN CYLINDER 'A' ROLL DOWN TO SET

DEVELOPMENT OF CYLINDER A WITH HOLE (BEFORE ROLLING)

Exercises for developing cylinders

1. Complete the development of the cylinder with both ends cut at angles to centre line.

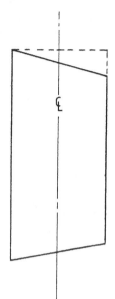

2. Complete the development of the cylinder *A* from the intersecting cylinders below.

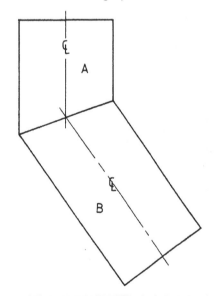

Exercises (cont'd)

3. Complete the following round branch on centre showing:
   (a) line of penetration;
   (b) development of branch B;
   (c) development of main cylinder A with hole.

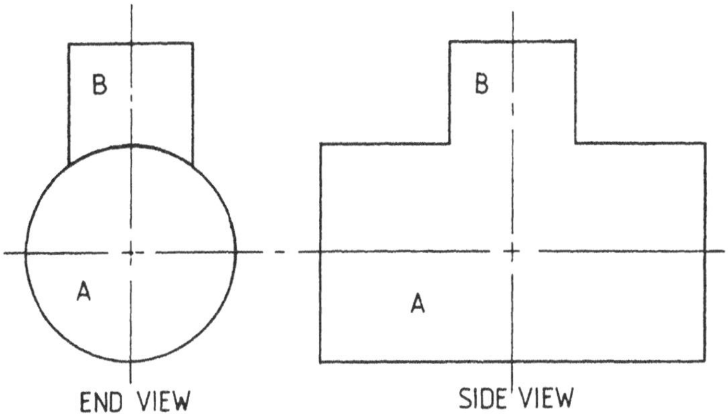

END VIEW

SIDE VIEW

Equal diameter T piece

Operation sequence

Front view

1. Draw in front view.
2. Draw in half top view of branch *B*, divide into six equal spaces and name *a* to *g*.
3. Draw in half end view of cylinder *A*, divide into six equal spaces and number 1 to 7.
4. Project points *b* to *f* and 2 to 6 parallel to centre lines to intersect on cylinder *A* and these intersection points form the line of penetration.
5. Draw in the line of penetration.

Development of branch B

1. Draw in rectangle, calculated mean circumference long by longest line *d* wide.
2. Divide length into twelve equal parts and label the points.
3. Project line of penetration points from front view on to corresponding lines on development.
4. Mark in the cut line with flexible straight edge to complete development of branch *B*.

Development of cylinder A with hole

1. Draw in rectangle, calculated mean circumference long by length of front view wide.
2. Divide length into twelve equal spaces and number the points.
3. Draw in the division lines numbered 4, 5, 6, 7, 6, 5, 4.
4. Project line of penetration points from front view on to corresponding lines of development to form cut out for hole.
5. Mark in line for hole using flexible straight edge to complete the development.

Calculation

$$MC = MD \times \pi$$
$$= 35 \times 3.1416$$
$$= 109.956$$
say, 110 mm

½ MC = 55 mm
¼ MC = 27.5 mm
$\frac{1}{12}$ MC = 9.16 mm

Development of equal diameter T piece

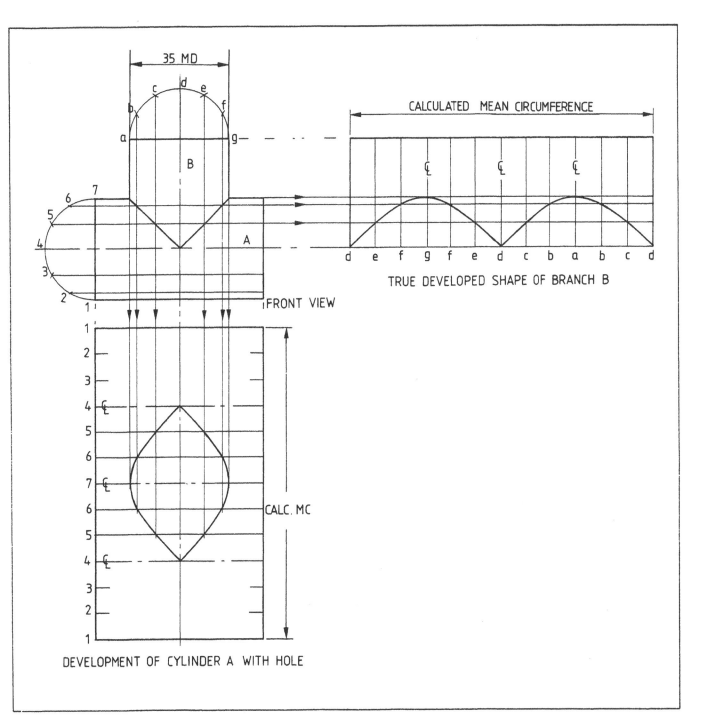

35 MD

B

A

FRONT VIEW

CALCULATED MEAN CIRCUMFERENCE

TRUE DEVELOPED SHAPE OF BRANCH B

d e f g f e d c b a b c d

CALC. MC

DEVELOPMENT OF CYLINDER A WITH HOLE

Round branch off centre and at 90° to centre line

Operation sequence

Front and end views
1. Lay out front view and end view.
2. Draw in half top view on branch pipe of both views.
3. Divide half top views into six equal parts.
4. Mark in development lines parallel to centre lines.
5. Number development lines.
6. Project points *A* to *M* from end view to corresponding lines of front view.
7. Draw lines through new points of front view to give the line of penetration.

Pattern for branch pipe B
1. Mark out rectangle, calculated circumference long by length *A*-1 wide.
2. Divide into twelve equal sections and number.
3. Transfer length of lines from end view on to the corresponding lines of pattern to locate points *A* to *M*.
4. Draw in developed shape with flexible straight edge to complete development of branch pipe *B*.

Pattern for cylinder A *with hole*
1. Mark out rectangle, calculated circumference long by length of front view wide.
2. Divide into two equal sections to locate centre line across plate.
3. Project development lines from the branch in front view on to plate for cylinder *A*.
4. Transfer points from end view (measured around the curve) to the plate for cylinder *A*.
5. The true shape of the hole is plotted through the intersecting points.

Calculations

MC of *A* = MD × π
   = 50 × 3.1416
   = 157.08 mm
½ MC = 78.54 mm

MC of *B* = MD × π
   = 40 × 3.1416
   = 125.664 mm
½ MC = 62.832 mm
¼ MC = 31.416 mm
$\frac{1}{12}$ MC = 10.472 mm

Development of round branch off centre and at 90° to centre line

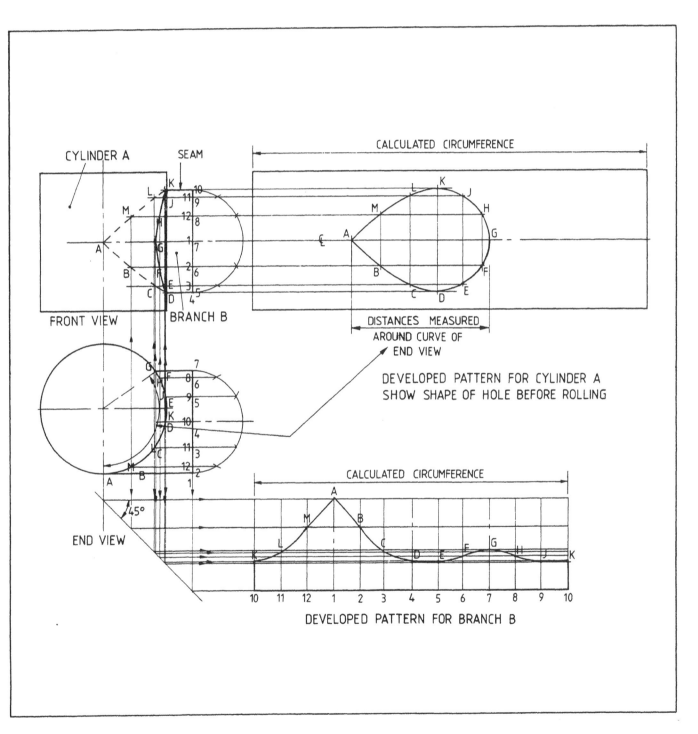

CYLINDER A

SEAM

CALCULATED CIRCUMFERENCE

FRONT VIEW

BRANCH B

DISTANCES MEASURED
AROUND CURVE OF
END VIEW

DEVELOPED PATTERN FOR CYLINDER A
SHOW SHAPE OF HOLE BEFORE ROLLING

END VIEW

45°

CALCULATED CIRCUMFERENCE

10 11 12 1 2 3 4 5 6 7 8 9 10

DEVELOPED PATTERN FOR BRANCH B

Connecting pipe on centre

Operation sequence

End and front view
1. Draw in outline of end view and front view.
2. Draw in parallel development lines from half end view of connecting pipe.
3. Draw in parallel development lines on front view of connecting pipe.
4. Project points from end view on to corresponding lines of front view to locate points for lines of penetration on both ends of the connecting pipe.
5. Draw lines of penetration on front view using flexible straight edge.
6. Draw in the position of *XY* line on both views.

Development of pattern for pipe B
1. Draw in rectangle, MC long by a width wider than lines 7 or 1 from the end view.
2. Divide rectangle into twelve equal parts and number the lines.
3. Draw in position of *XY* line.
4. Transfer lengths of development lines from above the *XY* line of front view to above the *XY* line of development to locate points *a* to *g* on the corresponding lines.
5. Transfer lengths of development lines from below the *XY* line of front view to below the *XY* line of development to locate points *h* to *n* on the corresponding lines.
6. Draw in both cut lines through development points using a flexible straight edge, to complete the development of connecting pipe *B*.

Calculation

$$MC = MD \times \pi$$
$$= 40 \times 3.1416$$
$$= 125.664 \text{ mm}$$
$$\text{say, } 126 \text{ mm}$$
$$\tfrac{1}{2} \text{ MC} = 63 \text{ mm}$$
$$\tfrac{1}{4} \text{ MC} = 31.5 \text{ mm}$$
$$\tfrac{1}{12} \text{ MC} = 10.5 \text{ mm}$$

Development of connecting pipe on centre

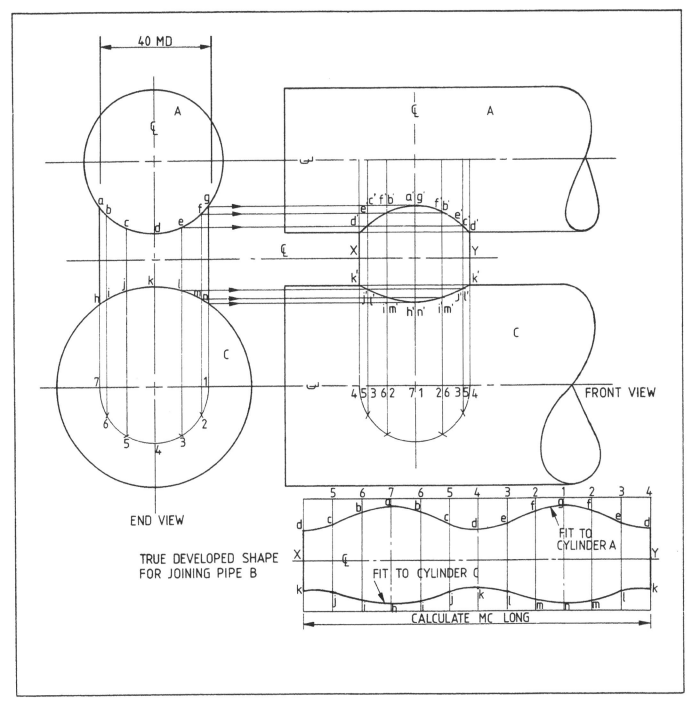

40 MD

A

END VIEW

FRONT VIEW

TRUE DEVELOPED SHAPE
FOR JOINING PIPE B

FIT TO
CYLINDER A

FIT TO CYLINDER C

CALCULATE MC LONG

Inclined branch — on centre (equal diameters)
Operation sequence

Front view

1. Draw in outline of front view.
2. Draw in half end view of branch A, divide into six equal parts and number 1 to 7.
3. Draw in half end view of main cylinder B, divide into six equal parts and number 1 to 7.
4. Project points 2 to 6 along front view parallel to centre line of cylinder B.
5. Project points 2 to 6 down parallel to centre line of branch B to intersect projected lines on cylinder B to locate points a to g which form the line of penetration.
6. Draw in the line of penetration to complete the front view.

Development of branch A

1. Draw in rectangle, mean circumference long and a distance longer than line 4d from front view wide.
2. Divide length into twelve equal spaces and number the points.
3. Transfer distance 1a from front view to lines 1 on development to locate points a'.
4. Transfer distance 2"b from front view to lines 2 on development to locate points b'.
5. Transfer distance 3"c from front view to lines 3 on development to locate points c'.
6. Transfer distance 4"d from front view to lines 4 on development to locate points d'.
7. Transfer distance 5"e from front view to lines 5 on development to locate points e'.
8. Transfer distance 6"f from front view to lines 6 on development to locate points f'.
9. Transfer distance 7g from front view to line 7 on development to locate point g'.

Development of main cylinder B

1. Draw in rectangle, mean circumference long by length of front view wide.
2. Divide length into twelve equal spaces and number the points.
3. Draw in lines 4, 5, 6, 7, 6, 5, 4 across the rectangle.
4. Project points a to g from front view down to intersect lines of development to locate points a" to g" to form shape of hole.

5. Draw in shape of hole through located points with flexible straight edge to complete development of cylinder A.

Calculation

$$MC = MD \times \pi$$
$$= 32 \times 3.1416$$
$$= 100.5312 \text{ mm}$$
$$\tfrac{1}{2} MC = 50.2656 \text{ mm}$$
$$\tfrac{1}{4} MC = 25.1328 \text{ mm}$$
$$\tfrac{1}{12} MC = 8.3776 \text{ mm}$$

Development of inclined branch — on centre (equal diameters)

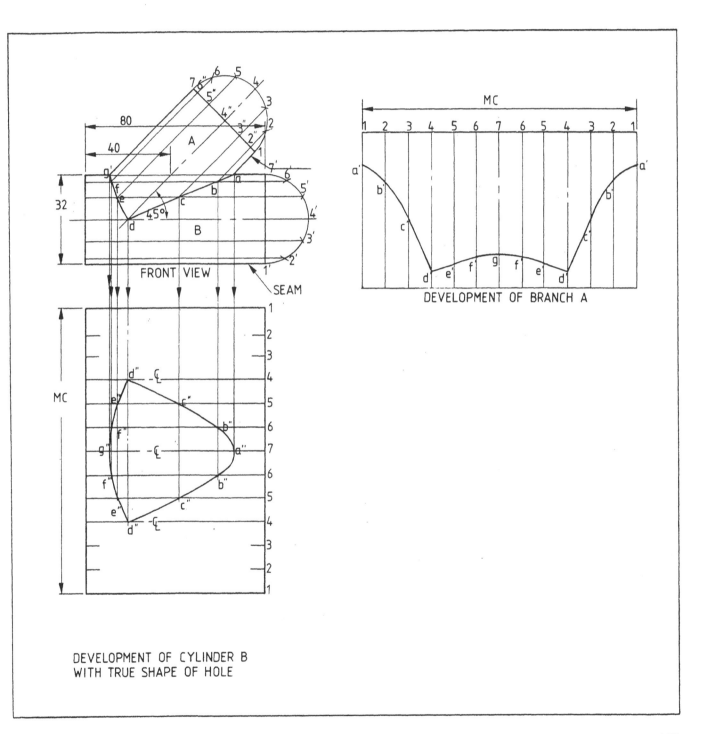

80

40

A

32

45°

B

d

e

f

g'

c

b

a

7'

6

5'

4'

3'

2'

1'

FRONT VIEW

SEAM

MC

1 2 3 4 5 6 7 6 5 4 3 2 1

a'

b'

c

d

e'

f'

g

f'

e

d'

c'

b'

a'

DEVELOPMENT OF BRANCH A

MC

1
2
3
4
5
6
7
6
5
4
3
2
1

d"
e"
f"
g"
f"
e"
d"

c"
c"
c"
b"
a"
b"
c"

DEVELOPMENT OF CYLINDER B
WITH TRUE SHAPE OF HOLE

197

Inclined branch — on centre (different diameters)

Operation sequence

Front and end views

1. Draw in outlines of front view and end view.
2. Draw in half end view of branch *F* in front view, divide into six equal spaces and project these points down parallel with the branch centre line.
3. Construct development lines on branch *F* in end view from half end view drawn above true opening.
4. Project points 1 to 12 from front view to end view to intersect development lines to give you true shape of the end of branch *F*.
5. Project points *A* to *M* across from end view to corresponding lines on front view to locate line of penetration.
6. Draw in line of penetration with a flexible straight edge.

Development of branch F

1. Mark in rectangle, MC long by any distance greater than length 10-*K* in front view.
2. Divide into twelve equal spaces and number the lines.
3. Transfer all lengths from front view 4*D*-3*C*-2*B*-1*A*-12*M*-11*L*-10*K*-9*J*-8*H*-7*G*-6*F* -5*E* to corresponding lines of development.
4. Draw in cut line with flexible straight edge to complete the development.

Development of hole in plate for cylinder E

1. Draw in rectangle, MC long by length of front view wide (if room permits).
2. Divide length in two to locate centre line.
3. Project up point *A* to *M* from line of penetration of front view to extend past the centre line.
4. Transfer points *A* to *M* measured around curve of end view and placed both sides of cross centre line of cylinder plate centre line.
5. The lines from steps 4 and 3 intersect to locate points *A* to *M* to form the true shape of the hole in cylinder before rolling.
6. Join points using flexible straight edge to complete the development.

Calculations

Cylinder E

$$MC = MD \times \pi$$
$$= 50 \times 3.1416$$
$$= 157.08 \text{ mm}$$
$$\tfrac{1}{2} \, MC = 78.54 \text{ mm}$$
$$\tfrac{1}{4} \, MC = 39.27 \text{ mm}$$
$$\tfrac{1}{12} \, MC = 13.09 \text{ mm}$$

Branch F

$$MC = MD \times \pi$$
$$= 40 \times 3.1416$$
$$= 125.664 \text{ mm}$$
$$\tfrac{1}{2} \, MC = 62.832 \text{ mm}$$
$$\tfrac{1}{4} \, MC = 31.416 \text{ mm}$$
$$\tfrac{1}{12} \, MC = 10.472 \text{ mm}$$

Development of inclined branch — on centre (different diameters)

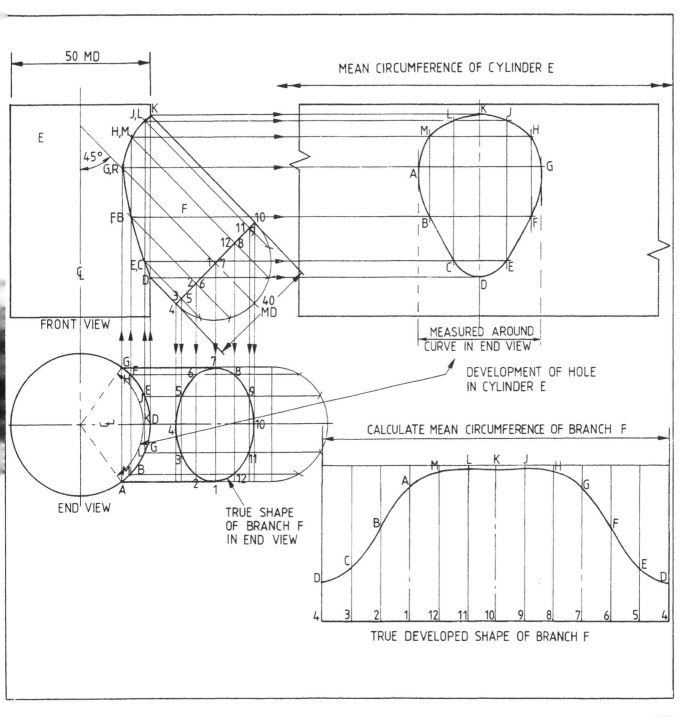

50 MD

MEAN CIRCUMFERENCE OF CYLINDER E

E

45°

FRONT VIEW

40 MD

MEASURED AROUND CURVE IN END VIEW

DEVELOPMENT OF HOLE IN CYLINDER E

END VIEW

TRUE SHAPE OF BRANCH F IN END VIEW

CALCULATE MEAN CIRCUMFERENCE OF BRANCH F

TRUE DEVELOPED SHAPE OF BRANCH F

Exercises for developing cylinders and connecting holes

1. Complete the offset branch pipe showing:
   (a) line of penetration;
   (b) developed shape of branch pipe;
   (c) developed shape of hole in main pipe.

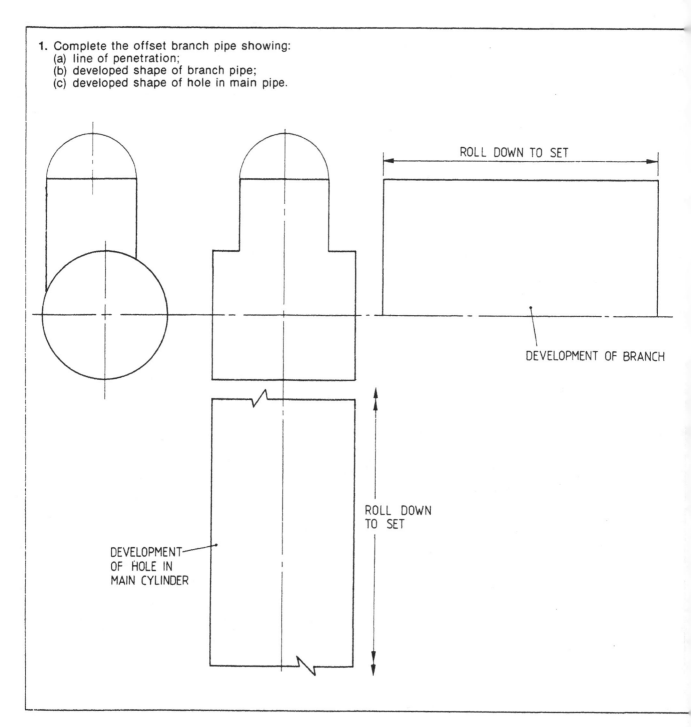

ROLL DOWN TO SET

DEVELOPMENT OF BRANCH

ROLL DOWN TO SET

DEVELOPMENT OF HOLE IN MAIN CYLINDER

2. Complete the inclined branch pipe showing:
   (a) line of penetration;
   (b) developed shape of branch pipe;
   (c) developed shape of hole in main pipe.

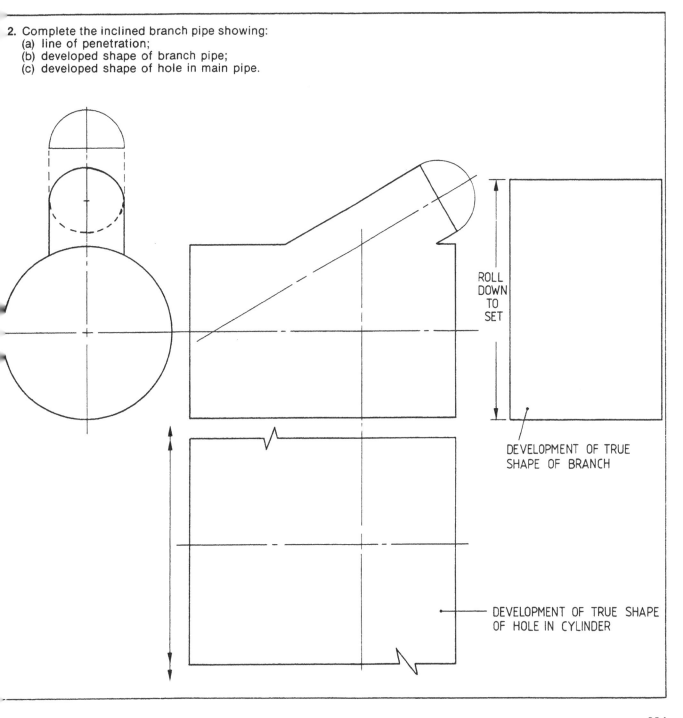

ROLL
DOWN
TO
SET

DEVELOPMENT OF TRUE
SHAPE OF BRANCH

DEVELOPMENT OF TRUE SHAPE
OF HOLE IN CYLINDER

Problem

To develop 90° bend made from four pieces.

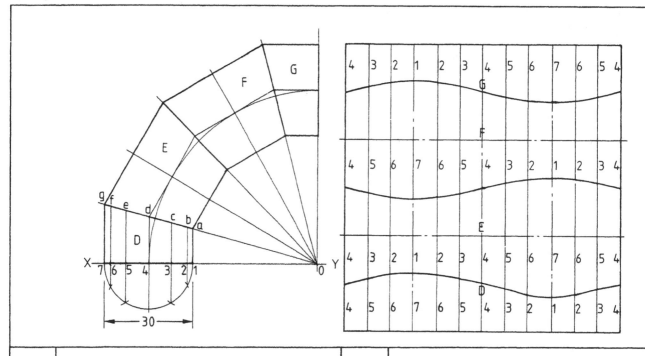

1   (a) Mark in end lines square to each other to locate point *O*.
      (b) Mark in centre lines to locate points *A*, *B* and *C*.
      (c) Calculate MC = MD × π
$$\begin{aligned} &= 30 \times 3.1416 \\ &= 94.248 \text{ mm} \end{aligned}$$
$$\begin{aligned} \tfrac{1}{2}\,MC &= 47.124 \text{ mm} \\ \tfrac{1}{4}\,MC &= 32.562 \text{ mm} \\ \tfrac{1}{12}\,MC &= 7.854 \text{ mm} \end{aligned}$$

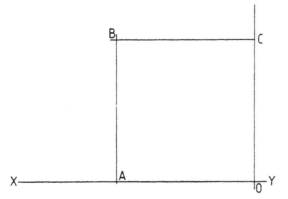

2   (a) Using *OA* as radius and point *O* as centre, scribe an arc.
      (b) Divide the arc into six equal sections.
      (c) Weld seam lines will be located at the 15°, 45°, 75° lines.
      (d) 30° and 60° will be the centre lines of the centre segments.

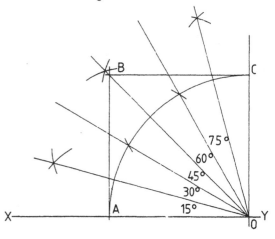

3	(a) Mark in half the MD each side of centre line point *A* on base line *XY* to locate points 1 and 7. (b) Square these points 1 and 7 up to the 15° weld seam line.	**4**	(a) Using *OZ* as radius and point *O* as centre, scribe arcs on to seam lines 45° and 75° to locate segment centre lines. (b) Join all these points to complete the centre lines of the segments.

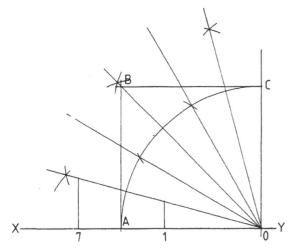

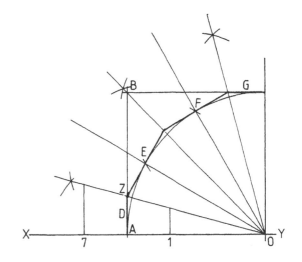

5	(a) Scribe half end view on to base line. (b) Divide half end view into say six equal sections. (c) Scribe points from half end view up parallel to centre line, to intersect the 15° weld seam line and to locate points 2, 3, 4, 5 and 6 on base line *XY*.	**6**	(a) Construct rectangle, MC long and height equal to sum of centre lines of segments *D, E, F, G*. (b) Divide length into twelve equal sections and number each line according to half end view in step 5. (c) Transfer lengths from front view on to corresponding lines of development plate. (d) If the rolls are wide enough, roll down in one piece and cut after rolling.

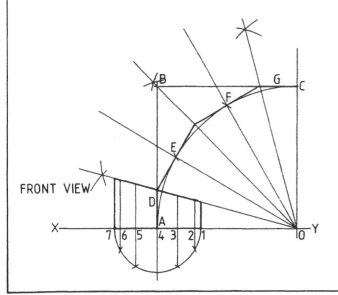

FRONT VIEW

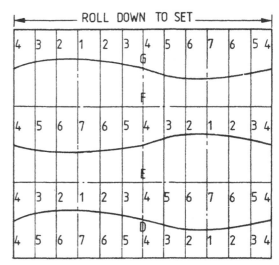

Vertical branch on 5-piece elbow
Operation sequence

Front view and half end view
1. Draw in 5-piece elbow.
2. Draw in position of branch.
3. Draw in half end view of branch.
4. Divide half end view into, say, six equal spaces.
5. Draw in development lines parallel to centre line of branch.
6. Number development lines 1 to 7.
7. Draw in half end view.
8. Draw in half top view of branch and divide into, say, three equal spaces.
9. Draw in lines from points and parallel to centre line of branch to locate points *a* to *g*.
10. Number lines to suit front view 1' to 7'.
11. Transfer points *a* to *g* from half end view to front view, keeping lines parallel to outside shape of front view to locate points *a'* to *g'*.
12. Draw in line of penetration through points *a'* to *g'*.

Development of shape for branch
1. Calculate MC of branch.

Calculation

$$MC = MD \times \pi$$
$$= 35 \times 3.1416$$
$$= 109.956 \text{ mm}$$
$$\tfrac{1}{2}\,MC = 54.978 \text{ mm}$$
$$\tfrac{1}{4}\,MC = 27.489 \text{ mm}$$
$$\tfrac{1}{12}\,MC = 9.163 \text{ mm}$$

2. Draw in rectangle, MC long by a width longer than line 7*g* from front view.
3. Divide length of rectangle into twelve equal spaces.
4. Number lines 1 to 7 to suit seam of branch in side view.
5. Transfer distances 1*a'*, 2*b'*, 3*c'*, 4*d'*, 5*e'*, 6 and 7*g'* to corresponding lines on development.
6. Draw in line through points *a'* to *g'* to *a'* with flexible straight edge to complete the developed pattern.

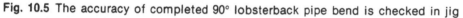

Fig. 10.5 The accuracy of completed 90° lobsterback pipe bend is checked in jig

Development of vertical branch on 5-piece elbow

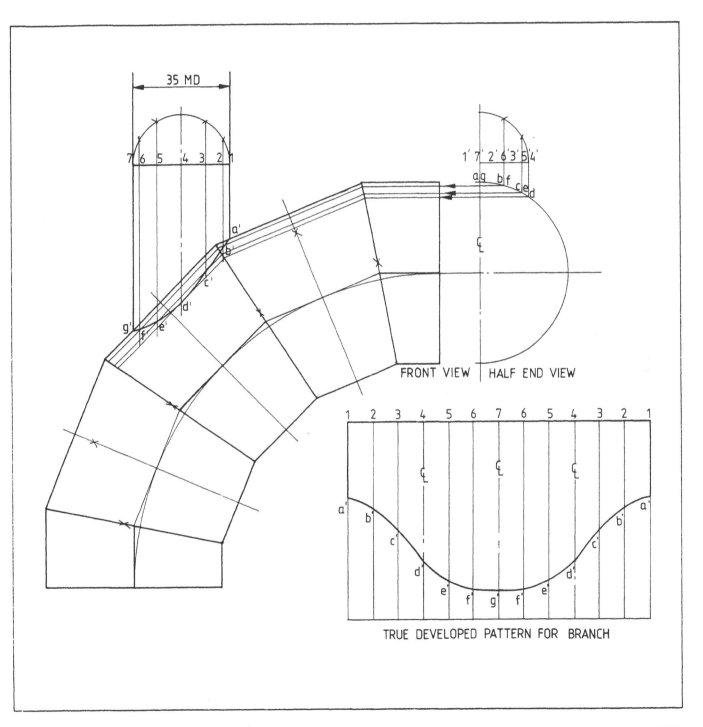

35 MD

7 6 5 4 3 2 1

a'
b'
c'
d'
g' f e'

1' 7' 2' 6' 3' 5' 4'
a g b f c e d
C̶L

FRONT VIEW | HALF END VIEW

1 2 3 4 5 6 7 6 5 4 3 2 1

C̶L   C̶L   C̶L

a'
b'
c'
d'
e' f' g' f' e' d'
c'
b'
a'

TRUE DEVELOPED PATTERN FOR BRANCH

Exercises for developing segmented bends with branch

1. Develop true pattern of 90° bend made up
   from four pieces A, B, C, D. Work from:
   (a) given bend radius 0–4;
   (b) given half end view.

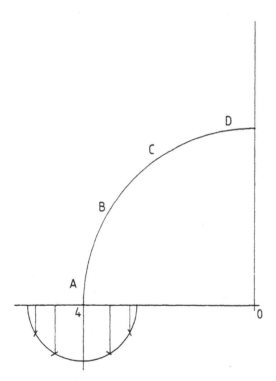

2. Develop true shape of vertical branch pipe (X).

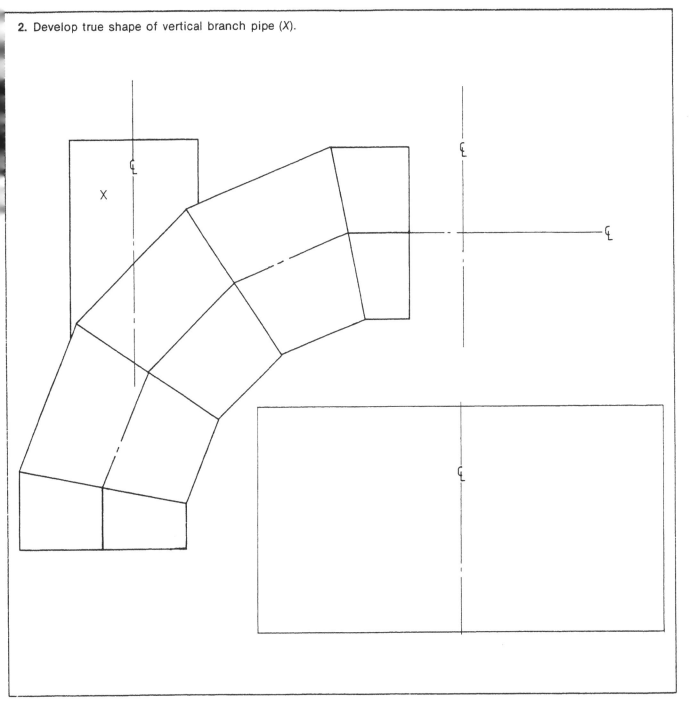

Square branch on centre to cylinder and at 90° to centre line

This job can be marked off in two ways.

Method 1

Using parallel lines in the end view. This method can be used when the main pipe can be either circular (large diam.) or elliptical and is shown on this page.

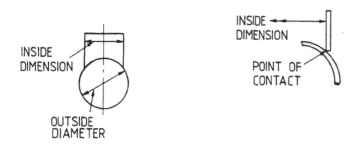

Divide end plate of square branch into four equal spaces. More spaces can be used if needed.

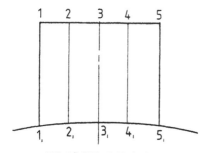

ENLARGED END VIEW

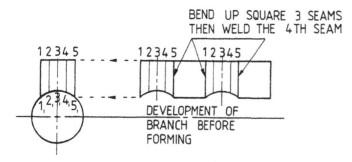

BEND UP SQUARE 3 SEAMS
THEN WELD THE 4TH SEAM

DEVELOPMENT OF
BRANCH BEFORE
FORMING

Square branch on centre to cylinder and at 90° to centre line (cont'd)

Method 2

Using dividers to swing a curve off the centre line. This method is used when the main pipe is circular only. Main methods use OD for main pipe and inside dimensions for the square branch.

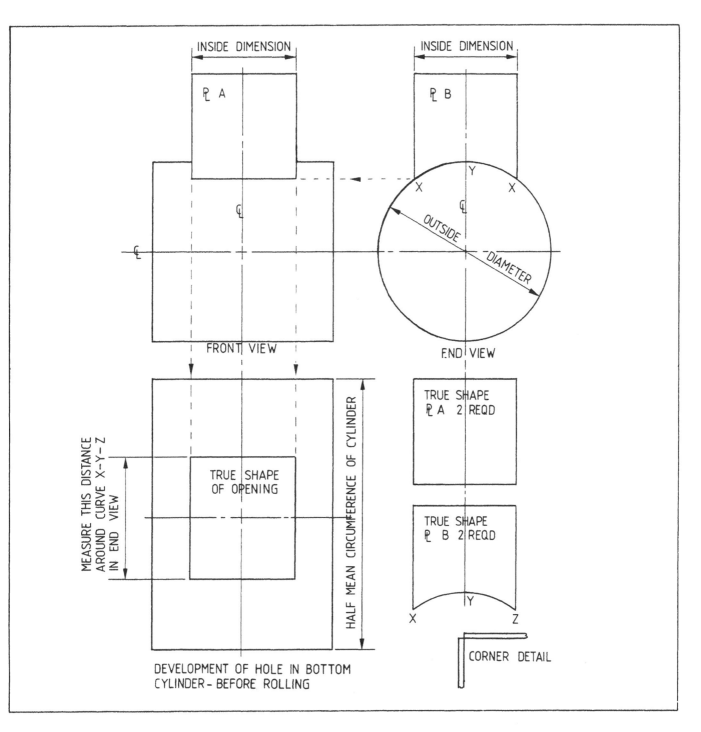

INSIDE DIMENSION

Pℒ A

INSIDE DIMENSION

Pℒ B

Y

X   X

℄

OUTSIDE DIAMETER

℄

FRONT VIEW

END VIEW

MEASURE THIS DISTANCE AROUND CURVE X–Y–Z IN END VIEW

TRUE SHAPE OF OPENING

HALF MEAN CIRCUMFERENCE OF CYLINDER

TRUE SHAPE Pℒ A 2 REQD

TRUE SHAPE Pℒ B 2 REQD

X   Y   Z

DEVELOPMENT OF HOLE IN BOTTOM CYLINDER – BEFORE ROLLING

CORNER DETAIL

Square branch on centre and at an angle to the centre line

Operation sequence

Front and end views

1. Construct front view and end view using OD for main cylinder and inside measurements for the inclined branch.
2. Divide branch in end view into any number of equal spaces (say eight) and identify each division line.
3. Transfer intersection points from end view on to the front view and identify points.
4. Draw in the true shape of branch side plate C using measurements from side view.
5. Draw in the true shape of long branch plate B using true lengths taken from the front view and placed on the corresponding line of the development.
6. Draw in the true shape of the short branch plate A using true lengths taken from the front view and placed on the corresponding line of the development.

Development of main cylinder with hole

1. Draw in rectangle, MC long by length of fr view wide.
2. Mark in distances measured around cu from end view, and transferred around cen line of development of plate D.
3. Project down true length points from front view to intersect the development lir from step 2.
4. Where the lines from steps 2 and 3 inters are the points for marking in the true de oped shape of the hole in plate D prior forming.

Calculation

$$MC = MD \times \pi$$
$$= 35 \times 3.1416$$
$$= 109.956 \text{ mm}$$
say, 110 mm
½ MC = 55 mm
¼ MC = 27.5 mm

Fig. 10.6 Inclined square branch on centre to cylinder

Development of square branch on centre and at an angle to the centre line

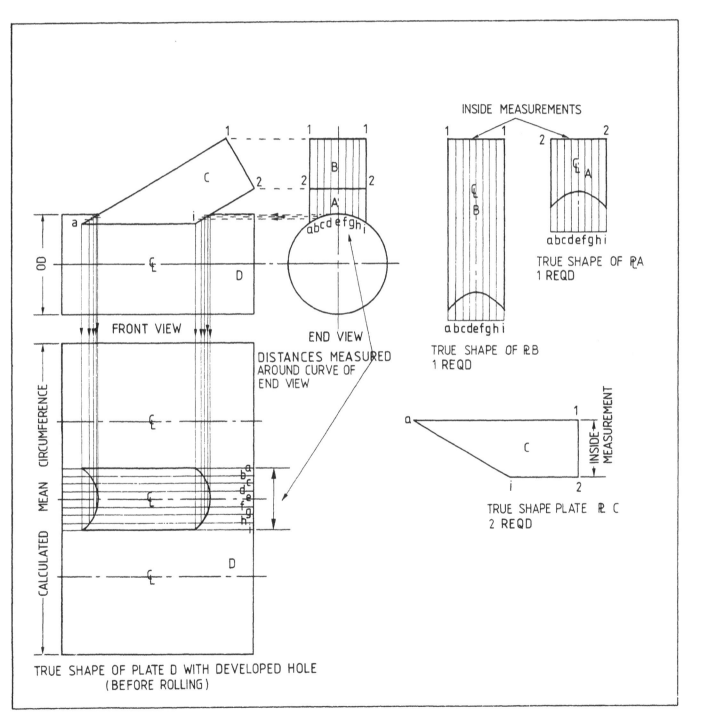

INSIDE MEASUREMENTS

B

A'

abcdefghi

END VIEW

DISTANCES MEASURED
AROUND CURVE OF
END VIEW

FRONT VIEW

OD

C

D

1

2

2

1

2

a

i

TRUE SHAPE OF Pℓ.A
1 REQD

abcdefghi

C
A

1

2

TRUE SHAPE OF Pℓ.B
1 REQD

abcdefghi

C
B

1

1

2

abcdefghi

CALCULATED   MEAN   CIRCUMFERENCE

a
b
c
d
e
f
g
h
i

C

D

TRUE SHAPE OF PLATE D WITH DEVELOPED HOLE
(BEFORE ROLLING)

TRUE SHAPE PLATE Pℓ. C
2 REQD

a

i

C

1

2

INSIDE MEASUREMENT

Development of square branch inclined to cylinder on centre

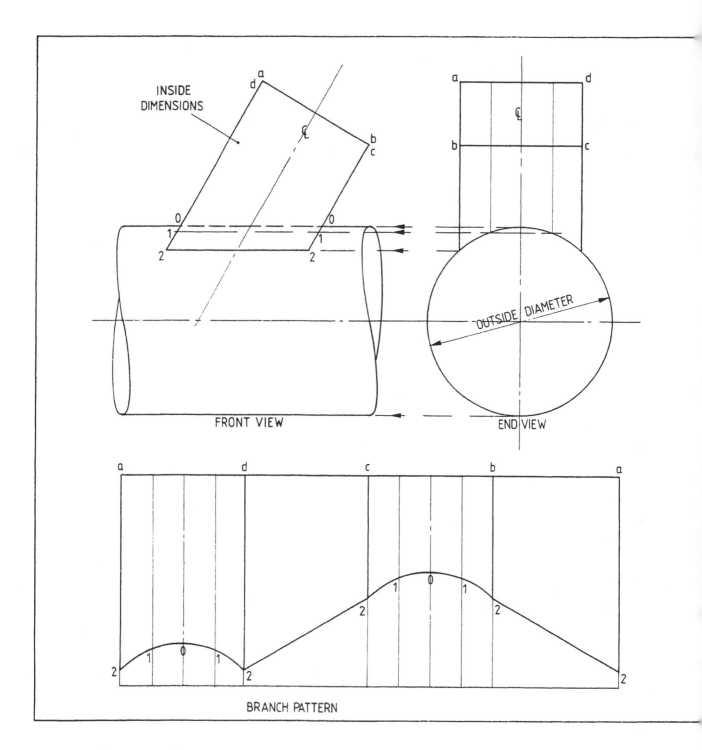

INSIDE DIMENSIONS

FRONT VIEW

END VIEW

OUTSIDE DIAMETER

BRANCH PATTERN

Notes

1. Develop as previous examples.
2. Use true lengths from front view for development of pattern.

212

Development of square branch inclined to cylinder diagonally fitted

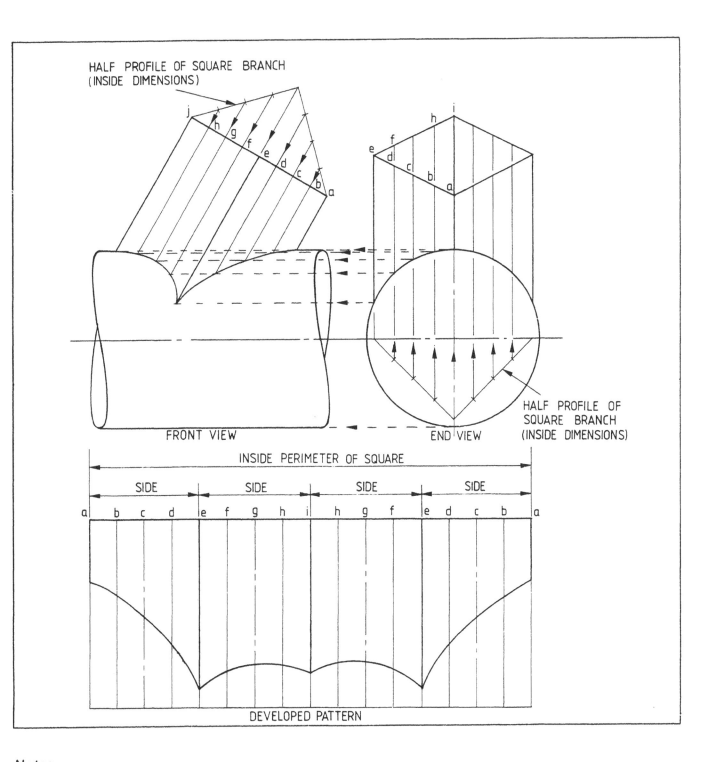

HALF PROFILE OF SQUARE BRANCH (INSIDE DIMENSIONS)

FRONT VIEW

END VIEW

HALF PROFILE OF SQUARE BRANCH (INSIDE DIMENSIONS)

INSIDE PERIMETER OF SQUARE

SIDE   SIDE   SIDE   SIDE

DEVELOPED PATTERN

Notes

1. Develop as previous examples.
2. Use true lengths from front view for development of pattern.

Exercises for developing square branches and connecting holes

1. Complete the following square branch on cylinder showing:
   (a) line of penetration;
   (b) development of plate *A* ⎱ corner to corner joints;
   (c) development of plate *B* ⎰
   (d) development of cylinder *C* with hole.

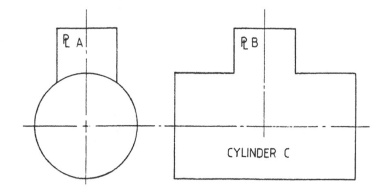

℄ A

℄ B

CYLINDER C

2. Complete the following inclined square
branch on centre to cylinder:
(a) line of penetration;
(b) development of plate A ⎤
(c) development of plate B ⎬ corner to corner joints;
(d) development of plate C ⎦
(e) development of half cylinder D with
hole.

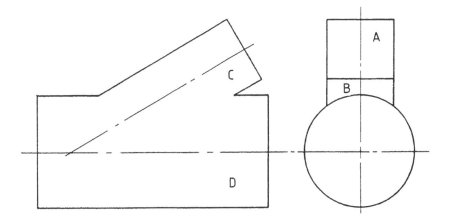

NOTES

11

Radial line development of conical surface

Laying out the true development pattern of any cone (where the apex is at a workable height) is most accurately marked out using radial line development.

Laying out the true developed pattern of right cones with inaccessible apex points or with apex points above workable height (slow tapered right cones) are best developed using:
1. two or three elevation methods;
2. verse sine method.

Laying out the true developed pattern of an oblique cone with an inaccessible apex point is by triangulation (see Ch. 13).

This chapter shows the development by radial line of:
1. right cone;
2. frustum of a right cone;
3. truncated right cone;
4. oblique cones;
5. frustum of oblique cones;
6. truncated oblique cones;
7. laying out using calculations;

followed by:
8. verse sine method to develop a frustum of a right cone;
9. two and three elevation methods to develop a frustum of a right cone;
10. radial line development of an equal tapered 90° bend.

Note

Do not measure from the diagrams except in exercises, because many figures are indication diagrams only and are not drawn to scale. All measurements on the diagrams are in millimetres unless stated otherwise.

Definitions

Right cone is one whose apex is over or co-incides in plan view with the centre of the base.
Frustum of a right cone is the base section of a right cone below a cutting plane which is parallel to the base.
Truncated right cone is the base section of a right cone below a cutting plane that is *not* parallel to the base.

Develop the pattern for a right cone by usin apex as centre and outside generator line (slar height) as radius, scribe a circular path, it distance being calculated mean circumferenc of base, then develop pattern between th circular path and apex.

Revise. Calculated mean circumference, i.e MC = MD × π (π = 3.1416). Make sure calc lated mean circumference is measured aroun curve.

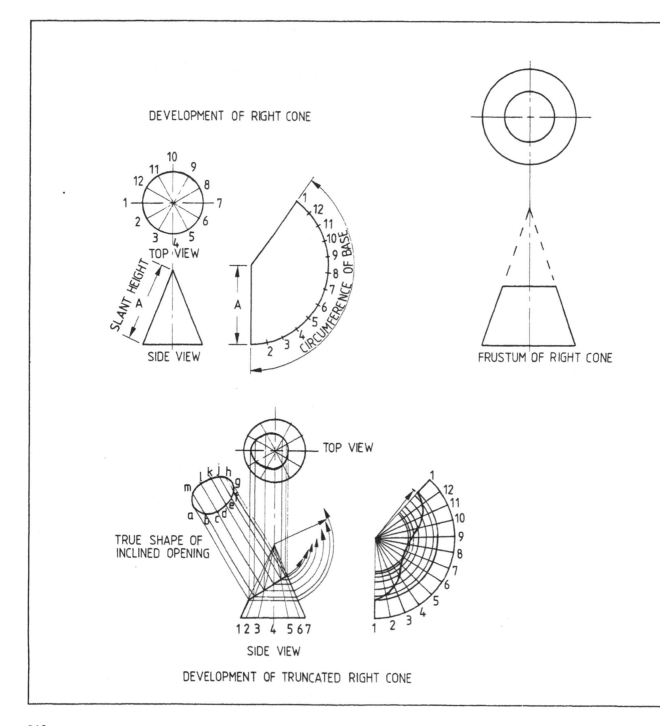

DEVELOPMENT OF RIGHT CONE

TOP VIEW

SLANT HEIGHT

SIDE VIEW

CIRCUMFERENCE OF BASE

FRUSTUM OF RIGHT CONE

TOP VIEW

TRUE SHAPE OF INCLINED OPENING

SIDE VIEW

DEVELOPMENT OF TRUNCATED RIGHT CONE

Problem

To develop a right cone.

Note Use mean dimensions.

Calculation

$$MC = MD \times \pi$$
$$= 40 \times 3.1416$$
$$= 125.664$$
$$\text{say, } 126 \text{ mm}$$

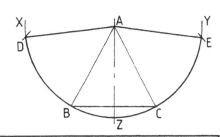

---

1 Draw in side view of cone *ABC* to mean dimensions.

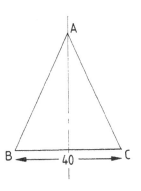

2 Using point *A* as centre and slope lengths *AB* and *AC* as radius, scribe a long arc *XY*, locating point *Z* on the centre line.

3
(a) From centre point *Z*, mark off around the arc half the mean circumference (63 mm) to locate point *D*.
(b) From centre point *Z*, mark off around the arc half the mean circumference (63 mm) to locate point *E*.

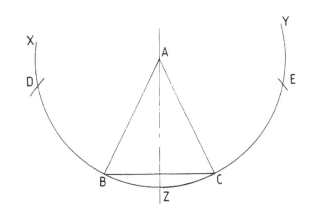

4
(a) Join points *D* and *E* to the apex point *A* to complete the developed pattern, *ADE*.
(b) Check the distance around the arc from point *D* to point *E* to make sure it is equal to the calculated MC (126 mm).

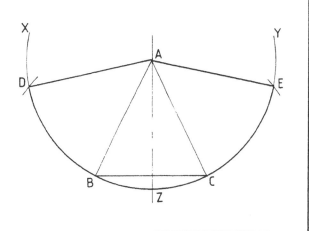

Problem

To develop frustum of a right cone.

Note Use mean measurements.

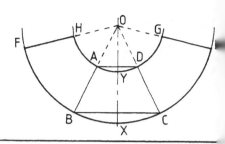

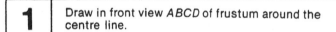

Calculations

H = apex height
h = vertical height of frustum
D = large mean diameter
d = small mean diameter

Apex height

$$H = \frac{D \times h}{D - d}$$

$$= \frac{40 \times 25}{40 - 15}$$

$$= \frac{1000}{25}$$

$$= 40 \text{ mm}$$

Large mean circumference

MC = MD × π
= 40 × 3.1416
= 125.664 mm
say, 126 mm
½ MC = 63 mm

Small mean circumference

MC = MD × π
= 15 × 3.1416
= 47.124 mm
say, 47 mm
½ MC = 23.5 mm

1	Draw in front view *ABCD* of frustum around the centre line.

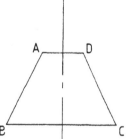

2	(a) Mark in calculated apex height of centre line to locate point *O*. (b) Join points *B*, *A* and *C*, *D* to apex point *O*.

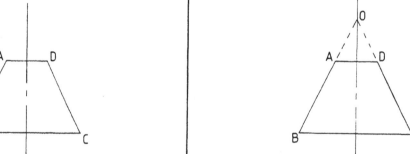

3	(a) Using point *O* as centre and *OB* and *OC* as radius, scribe an arc for the large end of developed pattern. (b) Using point *O* as centre and *OA* and *OD* as radius, scribe an arc for the small end of developed pattern.

4	(a) From centre point *X*, mark around the large curve half the calculated large MC to locate point *F*. (b) From centre point *X*, mark around the large curve half the calculated large MC to locate point *E*.

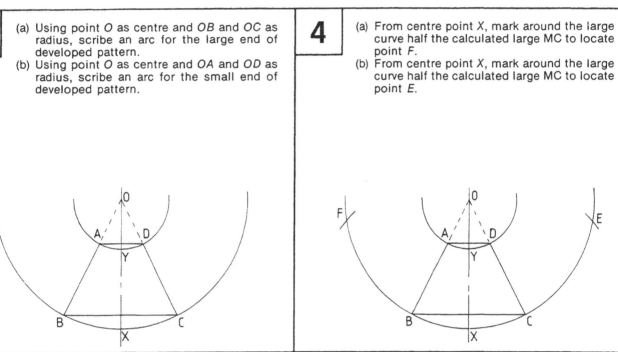

5	(a) Join point *F* to apex point *O* to locate point *H*. (b) Join point *E* to apex point *O* to locate point *G*. (c) Let *HFEG* be the true developed shape of the frustum.

6	*Checking technique* (a) Check point *F* to *E* around the curve to be equal to large calculated MC. (b) Check point *H* to *G* around the curve to be equal to small calculated MC. (c) Check that diagonals *EH* and *GF* are equal in length.

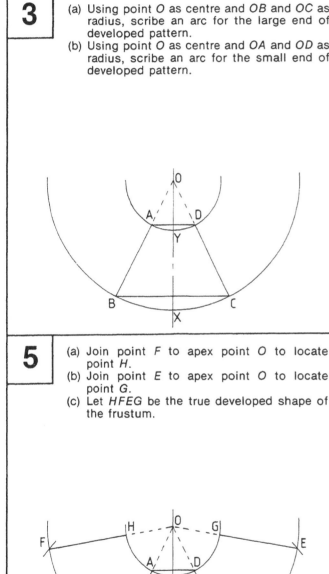

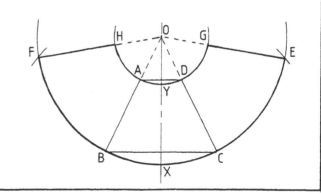

Problem
To develop a truncated right cone.

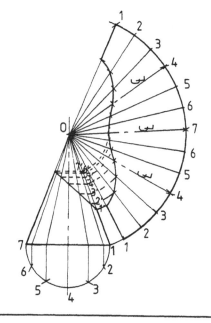

Calculations

Large mean circumference

$$MC = MD \times \pi$$
$$= 30 \times 3.1416$$
$$= 94.248 \text{ mm}$$
say, 94 mm
$$\tfrac{1}{2} MC = 47.124 \text{ mm}$$
$$\tfrac{1}{4} MC = 23.562 \text{ mm}$$
$$\tfrac{1}{12} MC = 7.85 \text{ mm}$$

Apex height

$$H = \frac{D \times h}{D - d}$$
$$= \frac{30 \times 25}{30 - 10}$$
$$= \frac{750}{20}$$
$$= 37.5 \text{ mm}$$

1
(a) Draw in front view *ABCD*.
(b) Mark in calculated apex height, point *O*.
(c) Extend side lines of front view to apex point *O*.

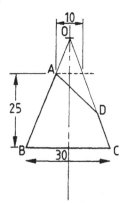

2
(a) Draw in half bottom view and divide into six equal spaces numbered 1 to 7.
(b) Project points 2 to 6 up parallel to centre line on to base line.
(c) Draw located points on base line up to apex point *O*.

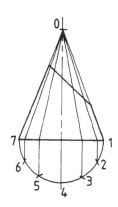

3	Square across the intersecting lines of the top line of the cone to meet the true length side line OC. Number these points 1' to 7' corresponding with the half bottom view.

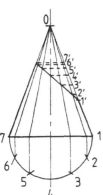

4	(a) Using point O as centre and OC as radius scribe in base curve.
	(b) Draw in a line from point O to base curve to locate start point 1".
	(c) From start point 1", mark in the calculated MC around the base curve to develop full cone.
	(d) Divide base curve into four equal sections to locate the centre lines.

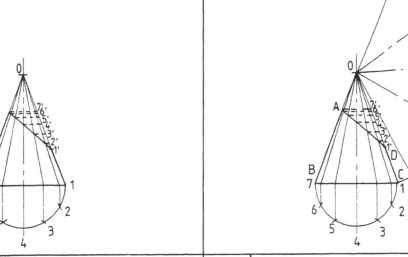

5	(a) Divide the base curve (calc. MC) into twelve equal spaces and number points according to half bottom view.
	(b) Join all these division points to apex point O.

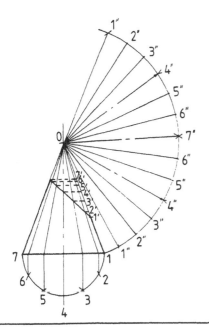

6	(a) Using apex point as centre, swing the points 1' to 7' on to the corresponding line of development to locate points 1" to 7".
	(b) Draw in top line with flexible curve to complete the developed shape of the truncated right cone.

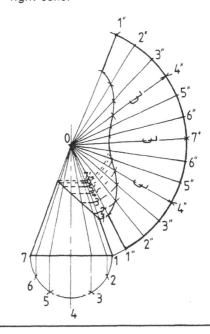

223

Frustum of right cones

By the use of calculations, all right cones and frustums of right cones can be developed without the use of a front view. Shown below are a series of calculations that would enable you to develop right cones without the front view being drawn.

Calculations for layout and development

Formulae

$$\text{Apex height, } H = \frac{D \times h}{(D - d)}$$

$$\text{Slant height, } SH = \sqrt{H^2 + R^2}$$
$$\text{Small circumference, } c = \pi d$$
$$\text{Large circumference, } C = \pi D$$
$$\text{Slant height of frustum, } sh = \sqrt{h^2 + (R - r)^2}$$

Example

Calculate the apex height, slant height, small and large circumference of a frustum of a right cone where:

D = 2700 mm
d = 950 mm
h = 1290 mm
R = 1350 mm
r = 475 mm

Calculations

$$
\begin{aligned}
H &= \frac{D \times h}{(D - d)} \\[4pt]
&= \frac{2700 \times 1290}{(2700 - 950)} \\[4pt]
&= \frac{3\,483\,000}{1750} \\[4pt]
&= 1990.285 \text{ mm}
\end{aligned}
$$

$$
\begin{aligned}
SH &= \sqrt{H^2 + R^2} \\
&= \sqrt{3\,961\,234 + 1\,822\,500} \\
&= \sqrt{5\,783\,734} \\
&= 2404.939 \text{ mm}
\end{aligned}
$$

$$
\begin{aligned}
sh &= \sqrt{h^2 + (R - r)^2} \\
&= \sqrt{1290^2 + (1350 - 475)^2} \\
&= \sqrt{1290^2 + 875^2} \\
&= \sqrt{1\,664\,100 + 765\,625} \\
&= \sqrt{2\,429\,725} \\
&= 1558.7575 \text{ mm} \\
&\quad \text{say, 1559 mm}
\end{aligned}
$$

$$
\begin{aligned}
C &= \pi D \\
&= 3.1416 \times 2700 \\
&= 8482.32 \text{ mm}
\end{aligned}
$$

$$
\begin{aligned}
c &= \pi d \\
&= 3.1416 \times 950 \\
&= 2984.52 \text{ mm}
\end{aligned}
$$

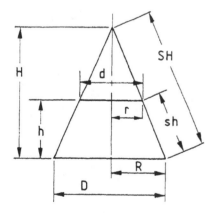

Exercise

Calculate the apex height, slant height, small and large circumference and slant height of a frustum of a right cone.

Calculations

$$H = \left(\frac{D \times h}{D - d}\right)$$

$$=$$

$$=$$

$$= \quad \text{mm}$$

$$SH = \sqrt{H^2 + R^2}$$

$$=$$

$$=$$

$$= \quad \text{mm}$$

$$sh = \sqrt{h^2 + (R - r)^2}$$

$$=$$

$$=$$

$$=$$

$$=$$

$$= \quad \text{mm}$$

$$C = \pi D$$

$$=$$

$$= \quad \text{mm}$$

$$c = \pi d$$

$$=$$

$$= \quad \text{mm}$$

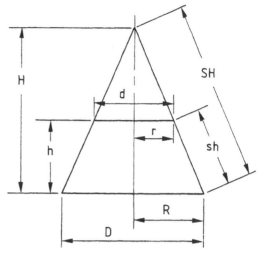

D = 1600 mm
d = 1100 mm
h = 900 mm
R = 800 mm
r = 550 mm

Problem

To develop a pattern of a frustum of a right cone using calculations and no front view.

Operation sequence

Calculations required
1. Calculate large end MD and MC.
2. Calculate small end MD and MC.
3. Calculate apex height (H).
4. Calculate slant height (SH) of the full cone.
5. Calculate slant height (sh) of the frustum.

Development of pattern
1. Mark in centre line.
2. Mark in base point B on the centre line.
3. Mark in slant height (1737 mm) from base point B to locate apex point A.
4. Mark in slant height of frustum (1023 mm) from base point B to locate point C.
5. Using apex point A as centre and distance AB as radius, scribe in the long base arc.
6. From point B, measure around base curve half the large MC (1162.5 mm) to locate points D and E. Double check length of curve DBE to be equal to large MC (2325 mm).
7. Join point D to A and point E to A.
8. Using apex point A as centre and distance AC as radius, scribe in the top arc to locate points F and G. Double check length of curve FCG to be equal to small MC (955 mm).
9. Check diagonal FE is equal in length to diagonal GD.
10. Firm in the completed pattern of the frustum, FCGEBD.

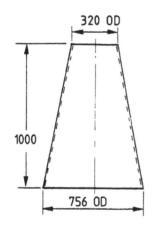

320 OD

1000

756 OD

MATERIAL: 16mm ℞

Calculations

$$\text{Large MD} = \text{OD} - \text{T}$$
$$= 756 - 16$$
$$= 740 \text{ mm}$$

$$\text{Large MC} = \text{MD} \times \pi$$
$$= 740 \times 3.1416$$
$$= 2324.784 \text{ mm}$$
$$\text{say, } 2325 \text{ mm}$$

$$\tfrac{1}{2} \text{ MC} = 1162.5 \text{ mm}$$

$$\text{Small MD} = \text{OD} - \text{T}$$
$$= 320 - 16$$
$$= 304 \text{ mm}$$

$$\text{Small MC} = \text{MD} \times \pi$$
$$= 304 \times 3.1416$$
$$= 955.0464 \text{ mm}$$
$$\text{say, } 955 \text{ mm}$$

$$\tfrac{1}{2} \text{ MC} = 477.5 \text{ mm}$$

$$H = \frac{D \times h}{D - d}$$
$$= \frac{740 \times 1000}{740 - 304}$$
$$= \frac{740\,000}{436}$$
$$= 1697.2477 \text{ mm}$$
$$\text{say, } 1697 \text{ mm}$$

$$SH = \sqrt{H^2 + R^2}$$
$$= \sqrt{1697^2 + 370^2}$$
$$= \sqrt{2\,879\,809 + 136\,900}$$
$$= \sqrt{3\,016\,709}$$
$$= 1736.8675 \text{ mm}$$
$$\text{say, } 1737 \text{ mm}$$

$$sh = \sqrt{h^2 + (R - r)^2}$$
$$= \sqrt{1000^2 + (370 - 152)^2}$$
$$= \sqrt{1000^2 + 218^2}$$
$$= \sqrt{1\,000\,000 + 47\,524}$$
$$= \sqrt{1\,047\,524}$$
$$= 1023.4861 \text{ mm}$$
$$\text{say, } 1023 \text{ mm}$$

**Developed pattern for frustum of
right cone using calculations and no
side view**

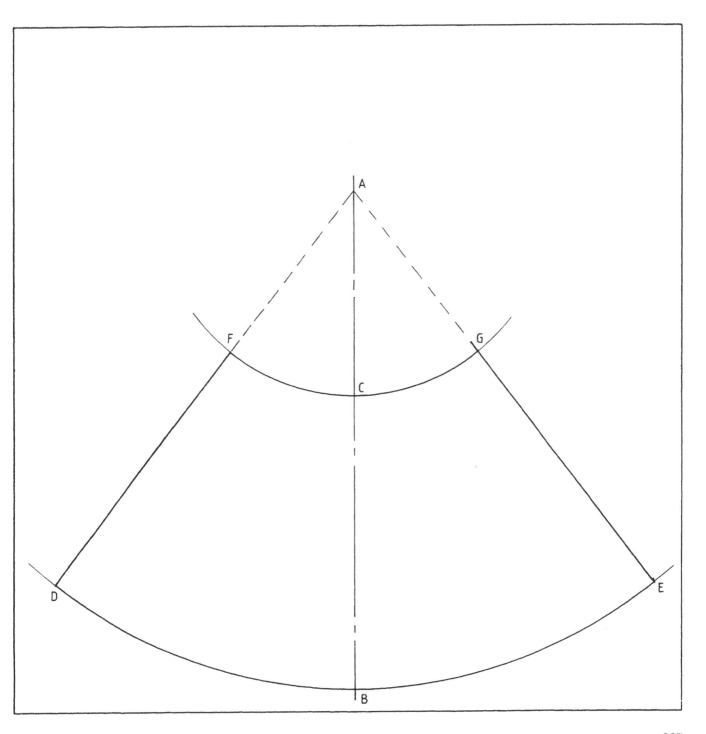

Exercises for developing cones

1. (a) Develop a full pattern of the right cone.
   (b) Calculate the large mean circumference.

2. (a) Develop full pattern of frustum of right cone without using a front view. Use a scale of 1:10.
   (b) Calculate apex point, slant height of base curve, slant height of top curve and both large and small MCs.

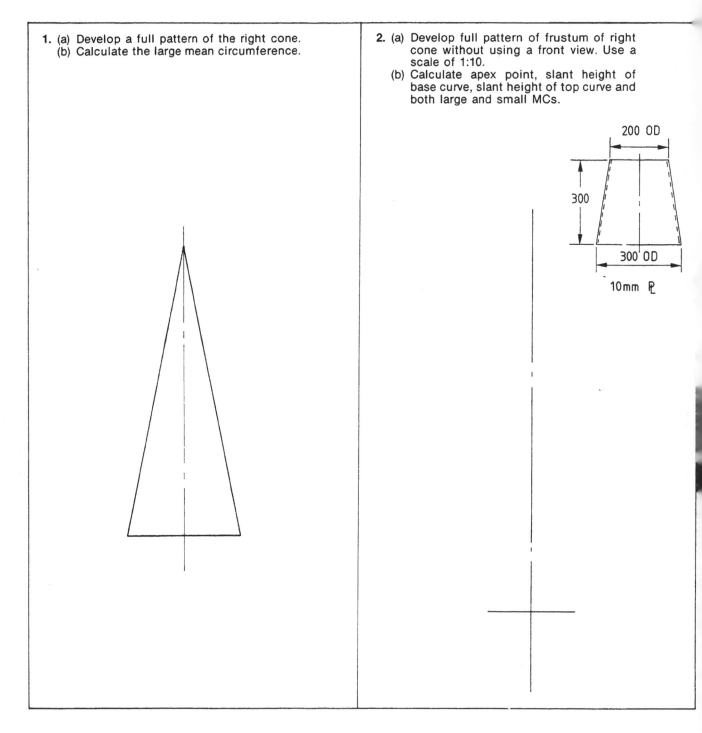

Exercises (cont'd)

3. (a) Develop a full pattern of the frustum of right cone.
   (b) Calculate apex height, small and large MC.

4. (a) Develop full pattern of truncated right cone.
   (b) Calculate apex height and large mean circumference.

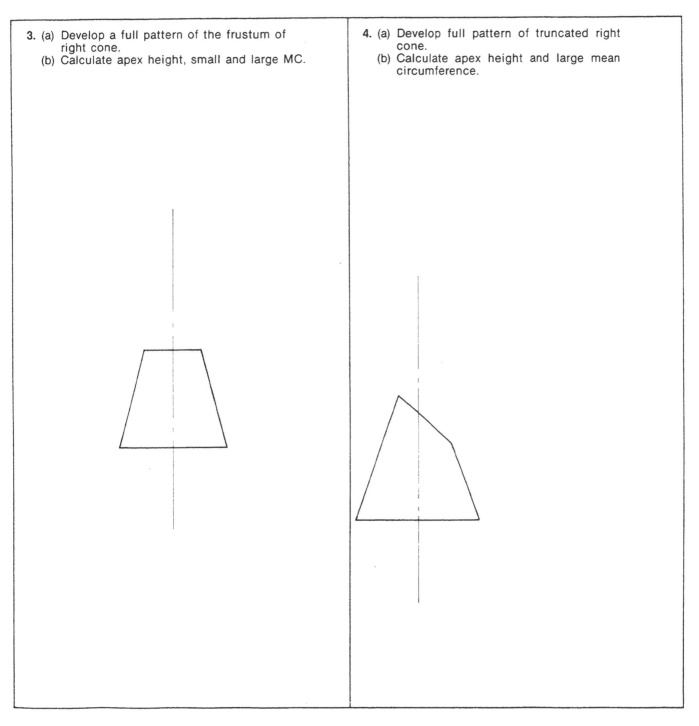

Practical example

Following is a handy job application showing the practical use of two different frustums of right cones (*A* and *B*), plus a truncated right cone (*C*), welded together to form a handy pouring jug. We suggest you make this from *3 mm aluminium plate*.

Calculations required for each cone are:
Small and large end MD and MC, plus apex height.

Note Use all mean dimensions for development.

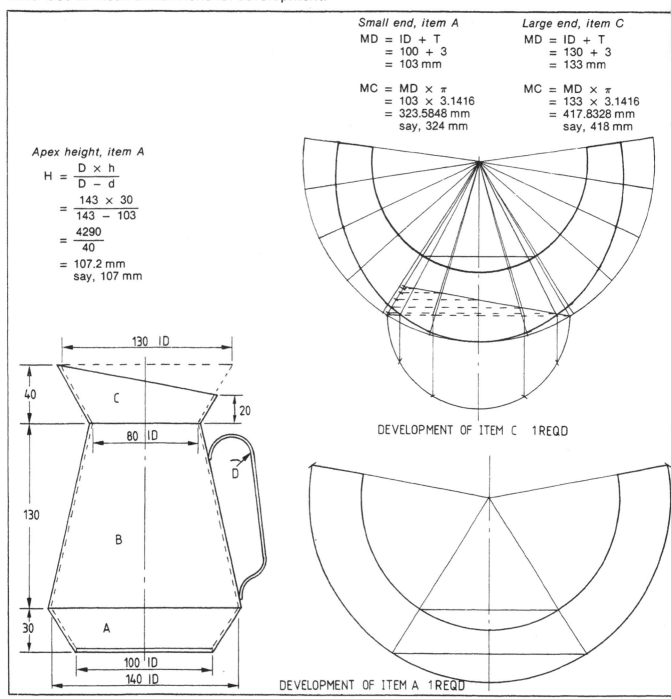

Small end, item A

MD = ID + T
   = 100 + 3
   = 103 mm

MC = MD × π
   = 103 × 3.1416
   = 323.5848 mm
   say, 324 mm

Large end, item C

MD = ID + T
   = 130 + 3
   = 133 mm

MC = MD × π
   = 133 × 3.1416
   = 417.8328 mm
   say, 418 mm

Apex height, item A

$$H = \frac{D \times h}{D - d}$$

$$= \frac{143 \times 30}{143 - 103}$$

$$= \frac{4290}{40}$$

= 107.2 mm
  say, 107 mm

DEVELOPMENT OF ITEM C   1REQD

DEVELOPMENT OF ITEM A 1REQD

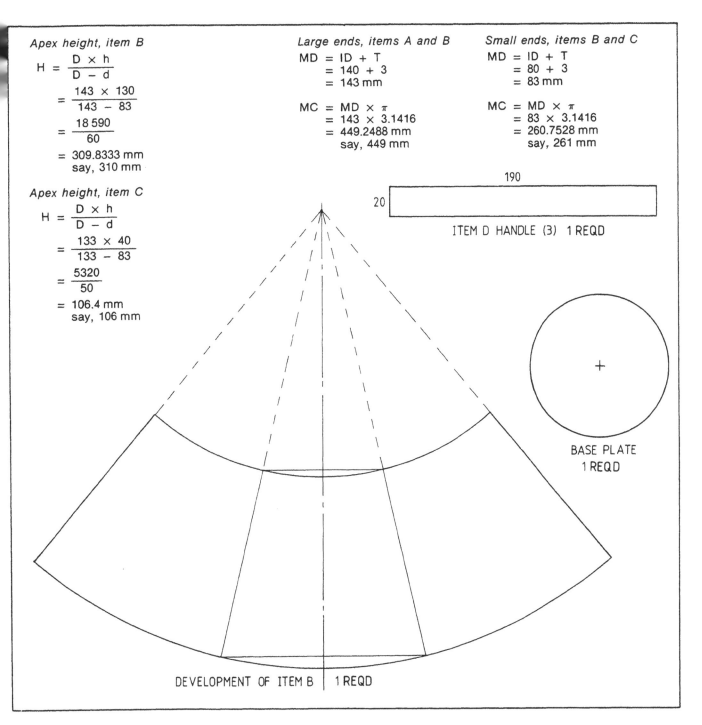

Apex height, item B

$$H = \frac{D \times h}{D - d}$$

$$= \frac{143 \times 130}{143 - 83}$$

$$= \frac{18\,590}{60}$$

$$= 309.8333 \text{ mm}$$
$$\text{say, } 310 \text{ mm}$$

Apex height, item C

$$H = \frac{D \times h}{D - d}$$

$$= \frac{133 \times 40}{133 - 83}$$

$$= \frac{5320}{50}$$

$$= 106.4 \text{ mm}$$
$$\text{say, } 106 \text{ mm}$$

Large ends, items A and B

$$MD = ID + T$$
$$= 140 + 3$$
$$= 143 \text{ mm}$$

$$MC = MD \times \pi$$
$$= 143 \times 3.1416$$
$$= 449.2488 \text{ mm}$$
$$\text{say, } 449 \text{ mm}$$

Small ends, items B and C

$$MD = ID + T$$
$$= 80 + 3$$
$$= 83 \text{ mm}$$

$$MC = MD \times \pi$$
$$= 83 \times 3.1416$$
$$= 260.7528 \text{ mm}$$
$$\text{say, } 261 \text{ mm}$$

190

20

ITEM D HANDLE (3) 1 REQD

BASE PLATE
1 REQD

DEVELOPMENT OF ITEM B | 1 REQD

Development of a slow tapering frustum of a right cone using verse sine and chordal lengths

Operation sequence
1. Draw in base line *XY*.
2. Construct centre line square to base line *XY*.
3. Mark in half calculated chordal length (1265 mm) each side of the centre line to locate points *A* and *B*.
4. Construct camber below base line to equal calculated verse sine distance of 66 mm (use a method from Chapter 8).
5. Check around developed camber line each side of centre line, point *C*, to equal half the calculated large mean circumference 1272.5 mm.

6. Using the calculated slant height (sh) of t
frustum as radius (900.5 mm), swing a seri
of arcs from say every second point of t
developed base camber.
7. Draw a curve line tangent to all the curv
using a flexible straight edge.
8. Measure around the top curved line ea
side of centre line point *D*, half the calc
lated small MC (1178 mm) to locate points
and *F*.
9. Join the points *A* to *E* and *B* to *F*.
10. Double check:
(a) Large calc. MC around curve *ACB*.
(b) Small calc. MC around curve *EDF*.
(c) Diagonal *EB* is equal to *AF*.
11. This is now the true developed shape of th
plate for the slow tapering frustum of a rig
cone, developed using verse sine and cho
dal lengths.

Explanation of signs
R = Large end radius
D = Mean diameter of base
d = Mean diameter of top
h = Vertical height of frustum
H = Apex height
sh = Slant height of frustum
SH = Slant height of right cone
VS = Verse sine height (height of base camber)
CL = Chordal length of base camber

Calculations

$$H = \frac{D \times h}{D - d}$$
$$= \frac{810 \times 900}{810 - 750}$$
$$= \frac{729\,000}{60}$$
$$= 12\,150 \text{ mm}$$
$$\text{or } 12.15 \text{ m}$$

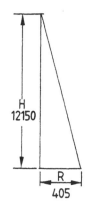

$$SH = \sqrt{H^2 + R^2}$$
$$= \sqrt{12.150^2 + 0.405^2}$$
$$= \sqrt{147.6225 + 0.164\,025}$$
$$= \sqrt{147.786\,52}$$
$$= 12.156\,748 \text{ m}$$
say, 12.157 metres
or 12 157 mm

MD of base = ID + T
         = 800 + 10
         = 810 mm

MC of base = MD × π
         = 810 × 3.1416
         = 2544.696 mm
         say, 2545 mm
         or 2.545 m
½ MC = 2.545 m ÷ 2
      = 1.2725 m
      or 1272.5 mm

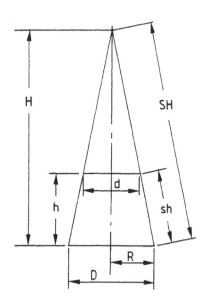

$$sh = \sqrt{h^2 + \left(\frac{D - d}{2}\right)^2}$$

$$= \sqrt{900^2 + \left(\frac{810 - 750}{2}\right)^2}$$

$$= \sqrt{810\,000 + \left(\frac{60}{2}\right)^2}$$

$$= \sqrt{810\,000 + 30^2}$$

$$= \sqrt{810\,000 + 900}$$

$$= \sqrt{810\,900}$$

$$= 900.499\,86 \text{ mm}$$
    say, 900.5 mm

$$\tfrac{1}{2}\,CL = \sqrt{[(2 \times SH) - VS] \times VS}$$

$$= \sqrt{[(2 \times 12\,157) - 66.58] \times 66.58}$$

$$= \sqrt{(24\,314 - 66.58) \times 66.58}$$

$$= \sqrt{24\,247.42 \times 66.58}$$

$$= \sqrt{1\,614\,393.2}$$

$$= 1270.5877 \text{ mm}$$
    say, 1271 mm
    or 1.271 m

$^{15}\!/_{64}$ of camber = 15.468 75 mm
$^{28}\!/_{64}$ of camber = 28.875 mm
$^{39}\!/_{64}$ of camber = 40.218 75 mm
$^{48}\!/_{64}$ of camber = 49.5 mm
$^{55}\!/_{64}$ of camber = 56.718 75 mm
$^{60}\!/_{64}$ of camber = 61.875 mm
$^{63}\!/_{64}$ of camber = 64.968 75 mm
$^{64}\!/_{64}$ of camber = 66 mm

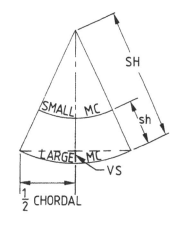

$$VS = \frac{D^2 \times 1.2337}{SH}$$

$$= \frac{810^2 \times 1.2337}{12\,157}$$

$$= \frac{809\,431}{12\,157}$$

$$= 66.581\,44 \text{ mm}$$

MD of top = ID + T
    = 740 + 10
    = 750 mm

MC of top = MD × π
    = 750 × 3.1416
    = 2356.2 mm
    say, 2356 mm
    or 2.356 m
½ of MC = 2.356 ÷ 2
    = 1.178 m
    or 1178 mm

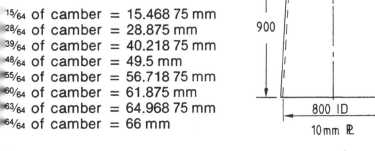

740 ID

900

800 ID

10 mm ℞

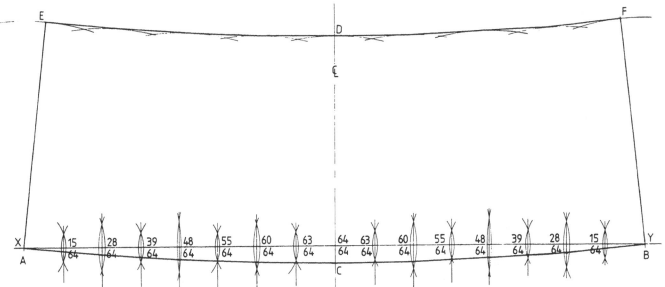

TRUE DEVELOPED SHAPE OF FRUSTUM

233

Three-elevation method used for developing true shape of plate required to form a slow tapering frustum of a right cone

Operation sequence

1. Draw in elevation *ABCD* using mean dimensions.
2. Draw in the other two elevations, each side of the first elevation, to locate points *E, F* and *G, H*.
3. Locate points *X, K* and *Y* as per enlarged sketch.
4. Using flexible straight edge draw in base curved line through points *H, X, A, K, B, Y, F*.
5. Locate points *Z, J* and *W* as per enlarged sketch.
6. Using flexible straight edge draw in curved line through points *G, Z, C, J, D, W*.
7. Each side of centre point *K*, measure around the curved base line half the calculated large MC (1272 mm) to locate points *L* and *M*.
8. Each side of centre point *J*, measure around the curved top line half the calculated small MC (1178 mm) to locate points *N* and *P*.
9. Join points *L* to *N* and *M* to *P* to complete the true developed shape of plate to form the slow tapering frustum of a right cone.
10. Double check:
    (a) Around base curve *LM* to equal large calc. MC.
    (b) Around top curve *NP* to equal small calc. MC.
    (c) Diagonal *NM* to be equal to diagonal *L*

Fig. 11.1 Frustum of a right cone

Calculations

Large end

$MD = OD - T$
$= 830 - 20$
$= 810 \text{ mm}$

$MC = MD \times \pi$
$= 810 \times 3.1416$
$= 2544.696 \text{ mm}$

$\tfrac{1}{2}MC = 2544.696 \div 2$
$= 1272.348 \text{ mm}$
say, 1272 mm

Small end

$MD = OD - T$
$= 770 - 20$
$= 750 \text{ mm}$

$MC = MD \times \pi$
$= 750 \times 3.1416$
$= 2356.2 \text{ mm}$

$\tfrac{1}{2}MC = 2356.2 \div 2$
$= 1178.1 \text{ mm}$
say, 1178 mm

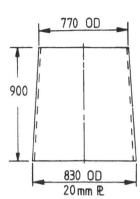

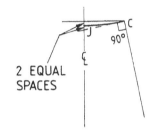

2 EQUAL SPACES

Enlarged view for locating points ZJW on top of centre line

Enlarged view for locating points XKY on bottom of centre line

2 EQUAL SPACES

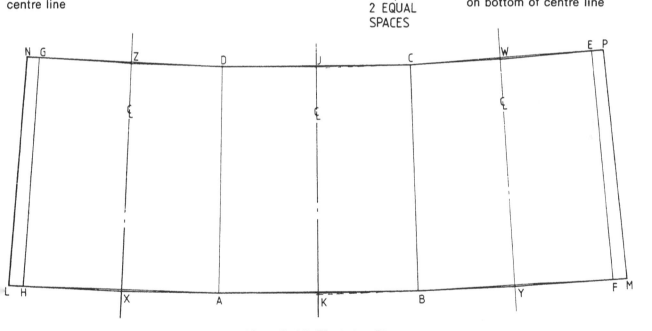

DEVELOPMENT OF TRUE SHAPE 1 ℞ REQD

Slow tapering right cone using the two-elevation method when the developed strake plate is a section of the circumference

Operation sequence

1. Draw in the two elevations, *CABD* and *AEFB* side by side, using mean dimensions.
2. Draw in straight line to join points *D* and *F* and to locate point *X* on centre line *AB*.
3. Distance *XB* will be the height of the camber distance for the base curved line.
4. Construct a camber using a method from Chapter 8.
5. Using slant height distance *AB* as radius and every second point of camber as centre, scribe a series of arcs.
6. Draw in the curved top line through points *C, A, E* and tangent to all the arcs scribed in step 5.
7. Using point *B* as centre, measure around the base curve one-sixth of large calc. MC (5760 mm) each side of the centre line *AB* to locate points *G* and *H*.
8. Using point *A* as centre, measure around the top curve one-sixth of the small calc. MC (5236 mm) each side of the centre line *AB* to locate points *J* and *K*.
9. Join points *G* to *J* and *H* to *K* to complete the developed shape of plate required.
10. Double check:
    (a) Base curve *GBH* to be equal to one-third of large MC.
    (b) Top curve *JAK* to be equal to one-third of small MC.
    (c) Diagonal *JH* to be equal in length to diagonal *GK*.

Calculations

Large MD

$$MD = OD - T$$
$$= 11\,030 - 30$$
$$= 11\,000 \text{ mm}$$
$$\text{or } 11 \text{ m}$$

Small MD

$$MD = OD - T$$
$$= 10\,030 - 30$$
$$= 10\,000 \text{ mm}$$

Large MC

$$MC = MD \times \pi$$
$$= 11\,000 \times 3.1416$$
$$= 34\,557.6 \text{ mm}$$
$$\text{say, } 34\,558 \text{ mm}$$
$$\tfrac{1}{3} MC = 34\,558 \div 3$$
$$= 11\,519 \text{ mm}$$
$$\tfrac{1}{6} MC = 5760 \text{ mm}$$

Small MC

$$MC = MD \times \pi$$
$$= 10\,000 \times 3.1416$$
$$= 31\,416 \text{ mm}$$
$$\tfrac{1}{3} MC = 31\,416 \div 3$$
$$= 10\,472 \text{ mm}$$
$$\tfrac{1}{6} MC = 5236 \text{ mm}$$

Camber calculations

$$\tfrac{15}{64} = 33.515\,625 \text{ mm}$$
$$\tfrac{28}{64} = 62.562\,5 \text{ mm}$$
$$\tfrac{39}{64} = 87.140\,625 \text{ mm}$$
$$\tfrac{48}{64} = 107.25 \text{ mm}$$
$$\tfrac{55}{64} = 122.890\,62 \text{ mm}$$
$$\tfrac{60}{64} = 134.062\,5 \text{ mm}$$
$$\tfrac{63}{64} = 140.765\,62 \text{ mm}$$
$$\tfrac{64}{64} = 143 \text{ mm}$$

Development of slow tapering right cone using the two-elevation method

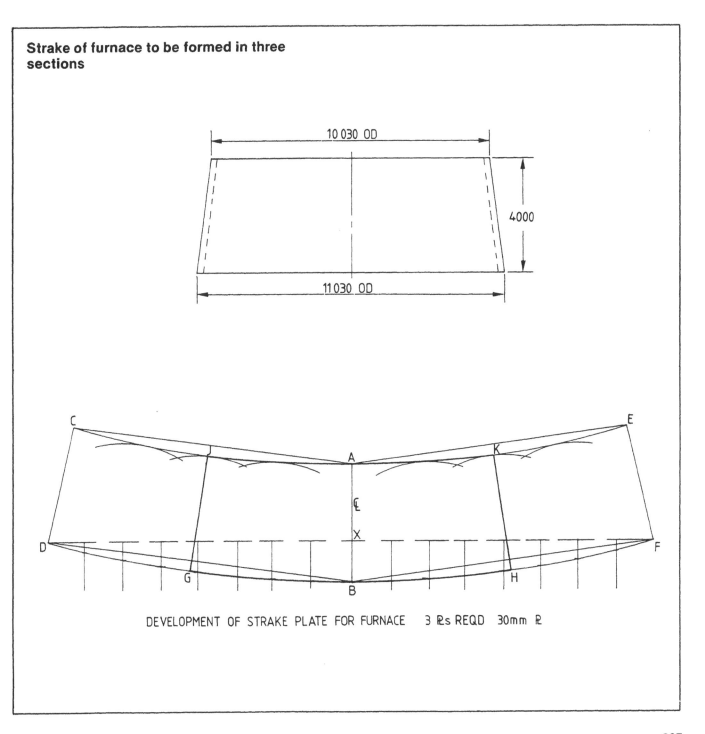

Strake of furnace to be formed in three sections

10 030 OD

4000

11 030 OD

DEVELOPMENT OF STRAKE PLATE FOR FURNACE   3 Ps REQD   30mm Ꝑ

Exercises for developing slow tapering cone frustums

1. Develop the full pattern of the frustum of the right cone as shown, using the verse sine method. Develop the pattern to a scale of 1:10.
   Calculate:
   (a) apex height (H);
   (b) slant height of cone (SH);
   (c) verse sine (VS);
   (d) MD and MC large end;
   (e) MD and MC of small end;
   (f) half chordal length;
   (g) slant height of frustum (sh);
   (h) camber heights.

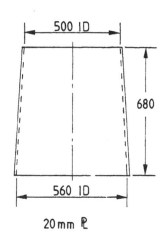

500 ID

680

560 ID

20 mm ℞

2. Develop the full pattern of the frustum of the right cone as shown, using the three-elevation method. Develop the pattern to a scale of 1:10.
   Calculate:
   (a) large end MD and MC;
   (b) small end MD and MC.

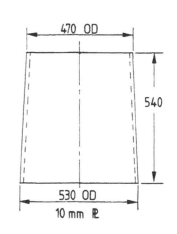

470 OD

540

530 OD

10 mm ℞

Oblique cones

Oblique cone is one whose apex is not over the centre of the base.

Frustum of an oblique cone is the base section below a cutting plane parallel to the base.

A cutting plane at an angle to the base is said to leave a "truncated" section from cutting plane to base.

Position seam on long side to give narrower shaped pattern to conserve plate. Check diagonals and circumferences for accuracy. Draw top and side views on mean measurements using information calculated, i.e. apex height (H) and offset distance (Y).

Apex height formula

$$H = \frac{D \times h}{D - d}$$

Horizontal distance

$$Y = \frac{D \times S}{D - d}$$

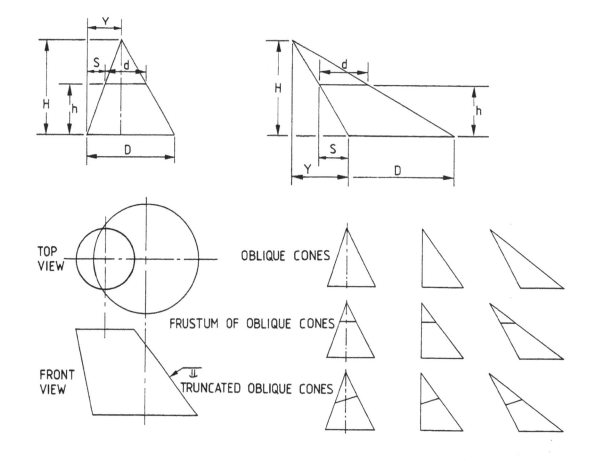

TOP VIEW

FRONT VIEW

OBLIQUE CONES

FRUSTUM OF OBLIQUE CONES

TRUNCATED OBLIQUE CONES

Position of seam on pattern

As mentioned on the previous page, the frustum of the oblique cone should have the weld seam placed on the long side of the cone.

This gives the developed pattern a longer narrower shape as shown with the black outline in contrast to the shorter wider pattern as shown with the thicker black outline (seam short side).

The longer, narrower pattern is easier to fit on to plate widths and makes the job easier to form in the press or the rolls.

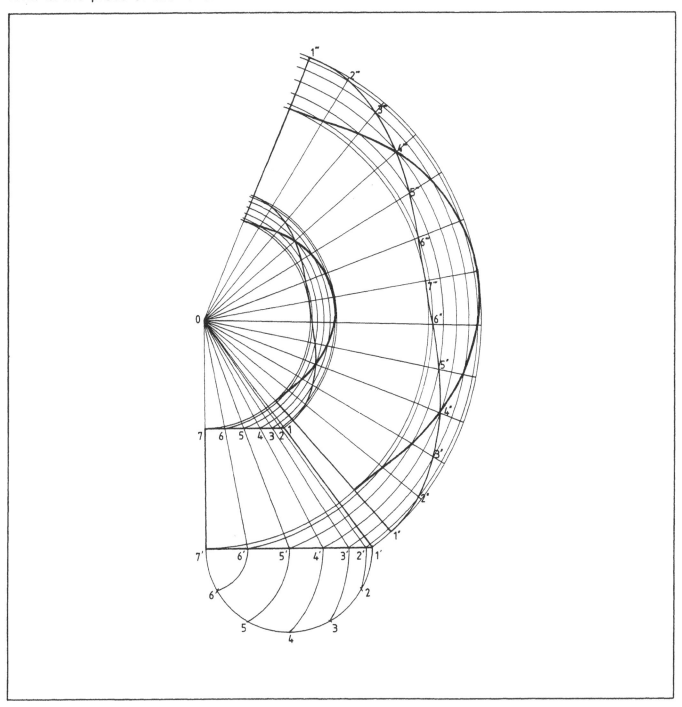

Problem

To develop an oblique cone frustum using radial line method (back corner obtuse angle, i.e. more than 90°).

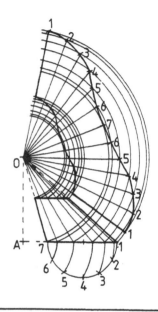

Calculations

Apex height

$$AH = \frac{D \times h}{D - d}$$

$$= \frac{25 \times 15}{25 - 12.5}$$

$$= \frac{375}{12.5}$$

$$= 30 \text{ mm}$$

Offset distance

$$Y = \frac{D \times S}{D - d}$$

$$= \frac{25 \times 4.5}{25 - 12.5}$$

$$= \frac{112.5}{12.5}$$

$$= 9 \text{ mm}$$

Large mean circumference

$$MC = MD \times \pi$$
$$= 25 \times 3.1416$$
$$= 78.54 \text{ mm}$$
$$\text{say, 79 mm}$$
$$\frac{1}{12} = 79 \div 12$$
$$= 6.5 \text{ mm}$$

Small mean circumference

$$MC = MD \times \pi$$
$$= 12.5 \times 3.1416$$
$$= 39.27 \text{ mm}$$
$$\text{say, 39 mm}$$

1
(a) Draw in front view of frustum to mean dimensions.
(b) Mark in calculated apex height.
(c) Mark in calculated apex point offset (Y distance) to locate point O.
(d) Extend side lines of side view to the apex point.

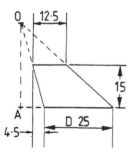

2
(a) Project apex point O drawn square to the baseline to locate point A.
(b) Draw in half bottom view of frustum, divide into six equal sections numbered 1 to 7.

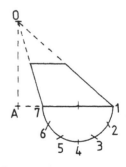

3	(a) Using point A as centre, scribe all points 2 to 6 around on to base line. (b) Join these points from base line up to apex point O.

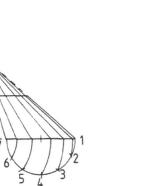

4	(a) Using apex point O as centre, swing all points 1 to 7 around to make long arcs. (b) Locate starting point for development, a line from apex point O to base arc line. (c) Using ¹⁄₁₂ of calculated large MC as radius, step off six equal spaces to locate points 2 to 7, stepping up one curved line each space. (d) Repeat (c) to locate other points 6 to 1, stepping down a curved line each space.

5	(a) Check around all base points of development to equal large calculated MC. (b) Join all located base points 1 to 7 to 1 to apex point O. (c) Join all located base points with flexible straight edge to complete development of base.

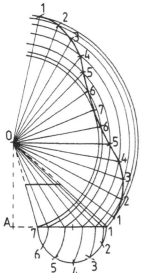
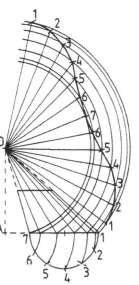

6	(a) Using apex point as centre, swing all points 1 to 7 of frustum top on to corresponding lines of the development to locate the points for the top curve. (b) Join all these located top points with flexible straight edge to complete the development of the frustum of the oblique cone (back, corner obtuse angle).

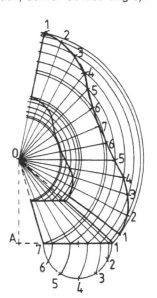

Radial line development: Frustum of an oblique cone back corner concave angle (less than 90°)

Operation sequence

1. (a) Draw in front view of frustum to mean dimensions.
   (b) Mark in calculated apex point height, 90 mm.
   (c) Mark in calculated apex offset distance (Y distance), 20 mm.
   (d) Extend side lines of side view to apex point.

2. (a) Project apex point O down square to base line of frustum to locate point a.
   (b) Draw in half bottom view of frustum.
   (c) Divide the half bottom view into six equal spaces and number the dividing points 1 to 7.

3. (a) Using point a as centre, swing all points 2 to 7 from half bottom view to base line of frustum.
   (b) Join all these new points on base line 7′ to 2′ up to apex point O, to make all these new lines true length lines.

4. (a) Using apex point O as centre, swing all points 7′ to 1′ from base line to form long curved lines.
   (b) Locate starting point, bring line from point O to any position on the long base curved line to position point 1″.
   (c) Using ¹⁄₁₂ of large MC as radius, step off from starting point 1″, six equal spaces, stepping up one curved line each space to locate points 2″, 3″, 4″, 5″, 6″, 7″.
   (d) Measure around located points 1″ to 7″ to equal half of large MC.

5. (a) Continue stepping out another six equal spaces, only step down a curved line each step to locate points 6″, 5″, 4″, 3″, 2″, 1‴.
   (b) Check measure around all base points to equal calculated MC.
   (c) Join all located points with a flexible straight edge to form the true developed shape of the bottom of the frustum.

6. (a) Join all points from developed shape of base, points 1″ to 7″ to 1‴ to apex point O.
   (b) Using apex point O as centre, swing all points 1′ to 7′ from top line of frustum, around on to the corresponding lines of the developed pattern.

(c) Join all located points to form true shape of top of developed cone and this completes the true developed shape of the frustum of the oblique cone.

Calculations

Apex height

$$H = \frac{D \times h}{D - d}$$

$$= \frac{70 \times 50}{70 - 31}$$

$$= \frac{3500}{39}$$

$$= 89.7 \text{ mm}$$

Offset distance

$$Y = \frac{D \times S}{D - d}$$

$$= \frac{70 \times 11.25}{70 - 31}$$

$$= \frac{787.5}{39}$$

$$= 20.1 \text{ mm}$$

Large mean circumference

$$MC = MD \times \pi$$
$$= 70 \times 3.1416$$
$$= 219.90 \text{ mm}$$
$$^{1}\!/_{12} \, MC = 18.3 \text{ mm}$$
$$^{1}\!/_{2} \, MC = 110 \text{ mm}$$

Small mean circumference

$$MC = MD \times \pi$$
$$= 31 \times 3.1416$$
$$= 97.3896 \text{ mm}$$
$$\text{say, } 97 \text{ mm}$$

Development of an oblique cone frustum (back corner less than 90°)

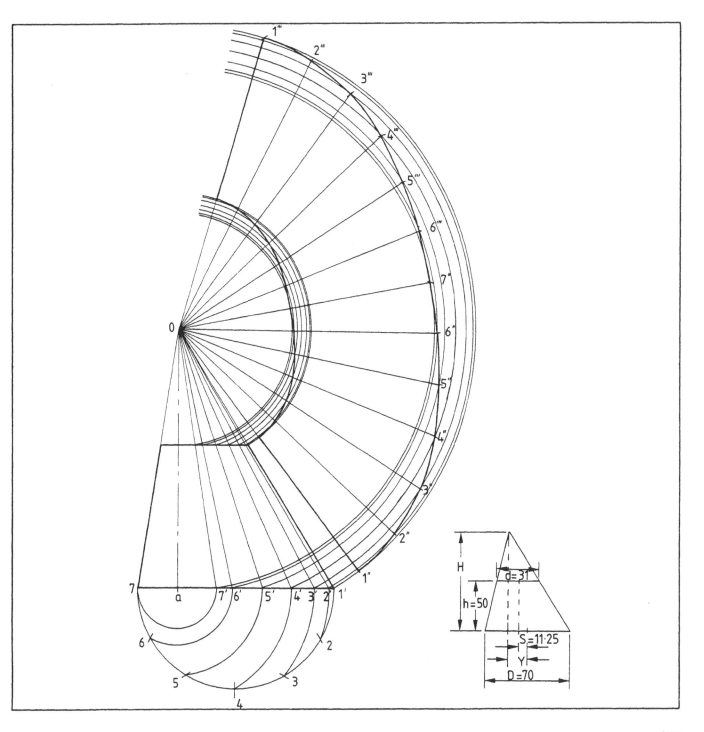

H

h = 50

d = 31

S = 11·25

Y

D = 70

Radial line development: Frustum of an oblique cone back corner 90° (straight back)

Operation sequence

1. (a) Draw in front view of frustum to mean dimensions.
   (b) Mark in calculated apex height, 96 mm.
   (c) Extend square back line of frustum to mean apex height.
   (d) Extend slope line of frustum to meet apex height and square line to locate point O.

2. (a) Draw in half bottom view of the frustum.
   (b) Divide the half bottom view of base into, say, six equal spaces.
   (c) Number these division points 1 to 7.

3. (a) Using point 7 as centre, swing points 2 to 6 up to locate points 2' to 6' on the base line.
   (b) Join the new points 2' to 6' from base line to apex point O and these lines are now true length lines.

4. (a) Using apex point O as centre, swing all points 1' to 7' from base line to form long curves.
   (b) Locate starting point, bring line from apex point O to any position on the long bottom curve.

5. (a) Using ¹⁄₁₂ of calculated large MC as radius, step off from starting point six equal spaces, stepping up a curved line each space to locate points 1″ to 7″.
   (b) Check measurement around points 1″ to 7″ to equal calculated MC.
   (c) Continue stepping another six equal distances as in step 5(a) but starting from point 7″ and stepping down a curved line each space to locate points 6‴ to 1‴.
   (d) Join all these located points using a flexible straight edge. This line then represents the true developed shape of the base of the frustum.
   (e) Check measure around plotted curve to make sure it is equal to the calculated large MC.

6. (a) Join all the plotted points 1″ to 7″ and 6‴ to 1‴ of the developed base line, to the apex point O.
   (b) Using apex point O as centre, swing all points 1 to 7 from top line of frustum (side view) around on to the corresponding lines of the developed pattern.

(c) Join all the located points using flexible straight edge to give you the true developed shape of the top of the frustum. This completes the developed shape of the straight back frustum of an oblique cone.

Calculations

Apex height

$$H = \frac{D \times h}{D - d}$$

$$= \frac{70 \times 50}{70 - 33.5}$$

$$= \frac{3500}{36.5}$$

$$= 95.89041 \text{ mm}$$
$$\text{say, } 96 \text{ mm}$$

Large mean circumference

$$MC = MD \times \pi$$
$$= 70 \times 3.1416$$
$$= 219.912 \text{ mm}$$
$$\text{say, } 220 \text{ mm}$$
$$^{1}\!/_{12} \text{ MC} = 220 \div 12$$
$$= 18.3 \text{ mm}$$
$$^{1}\!/_{2} \text{ MC} = 220 \div 2$$
$$= 110 \text{ mm}$$

Development of an oblique cone frustum (back corner 90°)

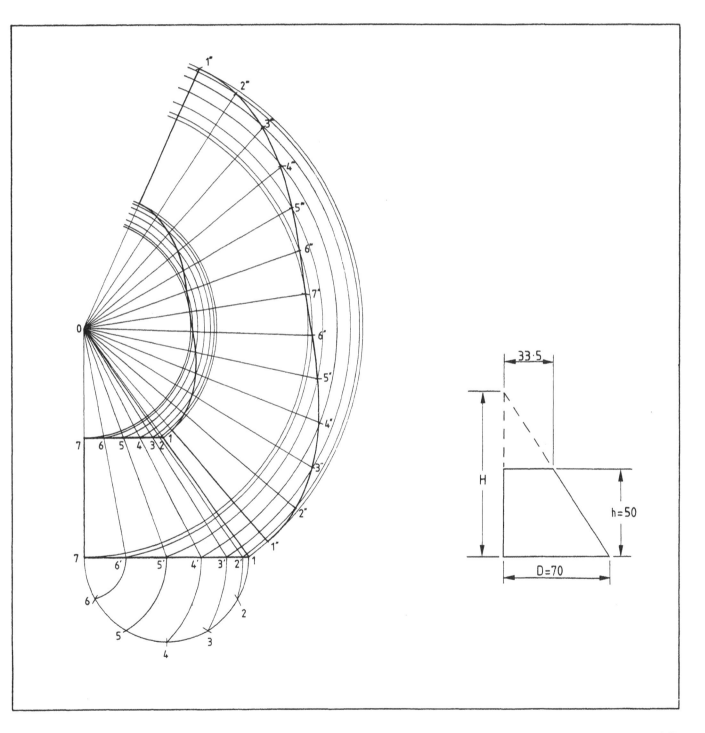

Radial line development: Truncated oblique cone

Operation sequence

1. (a) Draw in front view to mean dimensions.
   (b) Mark in calculated apex height, 63 mm.
   (c) Mark in calculated offset distance (Y), 27 mm, to locate point A.
   (d) Project up from point A, square to base line, to locate apex point O.
   (e) Extend side lines of side view to meet apex point O.

2. (a) Draw in half bottom view of base of cone.
   (b) Divide the half bottom view into six equal spaces.
   (c) Number these points 1 to 7.
   (d) Project all points 1 to 7 up vertical from half bottom view to base line, then join them to the apex point.
   (e) Label all points on truncation line a to g.

3. (a) Using point A as centre, swing all points 1 to 7 from half bottom view to the base line to locate points 1' to 7' in the true length diagram.
   (b) Draw all these points 1' to 7' up to apex point O. These lines now become true lengths.
   (c) Project all points a to g across parallel to base line to locate points a' to g' on the true length diagram.

4. (a) Using apex point O as centre, swing all points 1' to 7' from base line to form long curved lines.
   (b) Locate starting point, draw a line from point O to any position on the base curve.
   (c) Using 1/12 of large MC as radius, step off six equal spaces from starting point, stepping up one curved line each space to locate points 1" to 7".
   (d) Check measure around all points 1" to 7" to equal half large MC.

5. (a) Using 1/12 of large MC as radius, step another six equal spaces from point 7", stepping down a line each space to locate points 6''' to 1'''.
   (b) Check measure around located points 1" to 7" to 1''' to equal calculated large MC.
   (c) Draw in the true developed base line through all located points.

6. (a) Draw all located points 1" to 7" to 1'''' o developed base line up to point O.
   (b) Using apex point O as centre, swing a points a to g from truncation line of tru length view to the corresponding lines o the development to locate points a" to g to a".
   (c) Join all located points a" to g" to a''' t form true developed shape of the top an this completes the developed shape o the truncated oblique cone.

Calculations

Apex height

$$H = \frac{D \times h}{D - d}$$

$$= \frac{50 \times 33}{50 - 24}$$

$$= \frac{1650}{26}$$

$$= 63.461 \text{ mm}$$

say, 63 mm

Offset distance

$$Y = \frac{D \times S}{D - d}$$

$$= \frac{50 \times 14}{50 - 24}$$

$$= \frac{700}{26}$$

$$= 26.9 \text{ mm}$$

say, 27 mm

Large mean circumference

$$MC = MD \times \pi$$
$$= 50 \times 3.1416$$
$$= 157.08 \text{ mm}$$
$$\tfrac{1}{12} MC = 157.08 \div 12$$
$$= 13.09 \text{ mm}$$
$$\tfrac{1}{2} MC = 157.08 \div 2$$
$$= 78.54 \text{ mm}$$

say, 79 mm

Development of truncated oblique cone

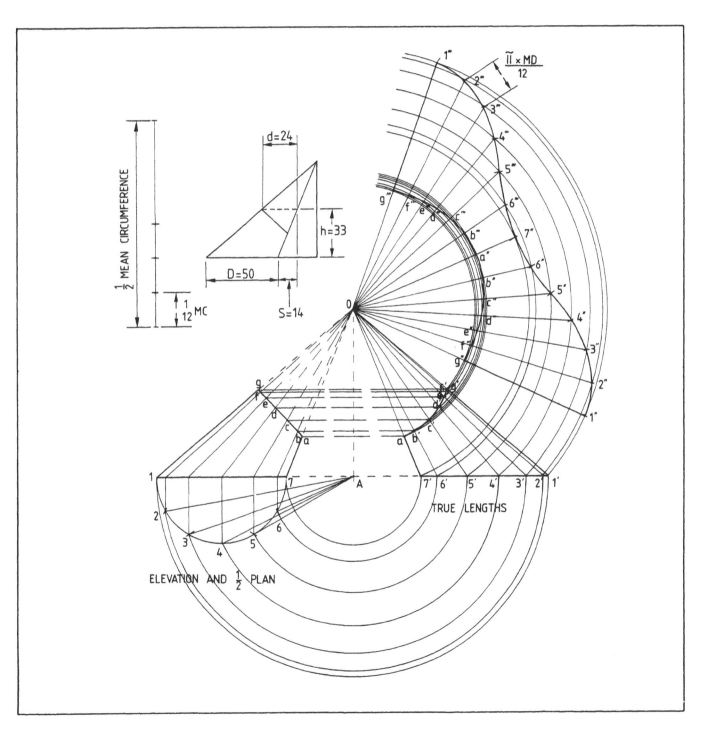

Exercises for developing eccentric conic sections

1. Develop full pattern frustum of oblique cone, where D = 31 mm, d = 15 mm, h = 27 mm. Calculate
   (a) apex height;
   (b) large MC;
   (c) small MC.

2. Develop full pattern frustum of oblique cone, where D = 27 mm, d = 15 mm, h = 23 mm, S = 4 mm. Calculate:
   (a) apex height;
   (b) offset distance Y;
   (c) large MC;
   (d) small MC.

3. Develop full pattern frustum of oblique cone, where D = 29 mm, d = 18 mm, h = 19 mm, S = 3 mm.
   Calculate:
   (a) apex height;
   (b) offset distance Y;
   (c) large MC;
   (d) small MC.

4. Develop full pattern truncated oblique cone where D = 19 mm, d = 8 mm, h = 26 mm, S = 4.5 mm.
   Calculate:
   (a) apex height;
   (b) offset distance Y;
   (c) large MC;
   (d) small MC.

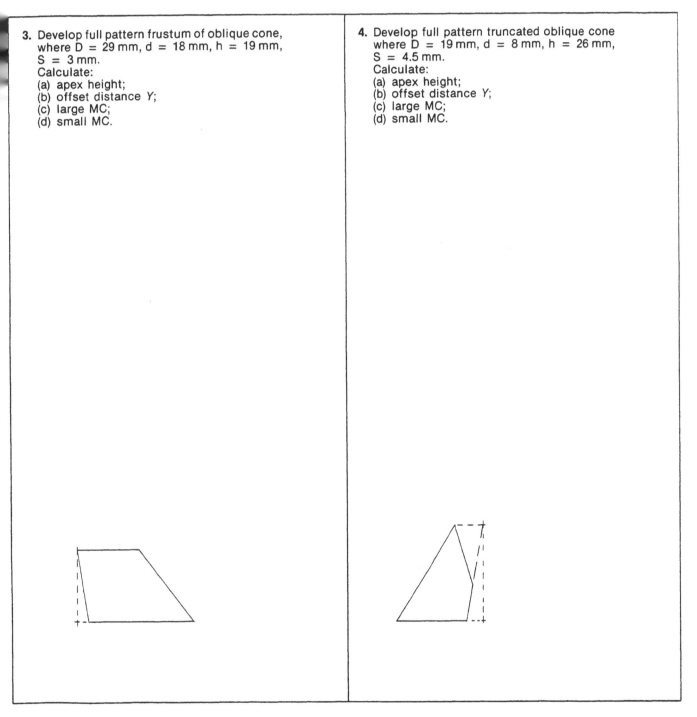

90° conical reducing elbow (equal tapering)

Operation sequence

Profile of bend

1. Draw in square end lines, locate point O at intersection.
2. Using point O as centre and given radius (500 mm), scribe in the arc AE.
3. Divide the arc AE into six equal spaces and number them 1 to 7.
4. Draw in seam lines 2-O, 4-O, 6-O which will be at 15°, 45°, 75° to the base line.
5. Square point A up to seam line to locate point B, square point E across to seam line to locate point D.
6. Using point O as centre and OB as radius, scribe points C and D on seam lines.
7. Join points B to C and tangent with point 5, then join points C to D and tangent with point 3.

   Note It will not be necessary to complete the full side profile because this is only to show the profile of the assembled bend.

Cone profile

1. Draw in frustum to mean dimensions: large end 460 mm, small end 230 mm, height = distance from profile of bend. Marked up the centre line of cone AB + BC + CD + DE + cut allowances for seams.
2. Draw in seam lines through points B, C, D at 15° to the horizontal.
3. Draw in half bottom view of base, divide into six equal parts, project these points up parallel to centre line to meet the base line, then join these points to apex point X.
4. Project points from seam lines across to the side slope line, parallel to the base line.

Development of shape

1. Using apex point X as centre and Xa as radius, scribe long base curve.
2. Locate starting point on base curve.
3. Measure large MC (1445 mm) around base curve from starting point.
4. Divide MC of base curve into twelve equal spaces and join the points from base curve to apex point X.

5. Using X as centre, scribe arcs from located points on slope line to development of shape.
6. Mark in cut lines as in truncated right cone (pp. 222–3) to complete the development.

Calculations

Large MC

$$MC = MD \times \pi$$
$$= 460 \times 3.1416$$
$$= 1445.136 \text{ mm}$$
$$\text{say, } 1445 \text{ mm}$$

Small MC

$$MC = MD \times \pi$$
$$= 230 \times 3.1416$$
$$= 722.568 \text{ mm}$$
$$\text{say, } 723 \text{ mm}$$

Apex height

$$H = \frac{D \times h}{D - d}$$
$$= \frac{460 \times 790}{460 - 230}$$
$$= \frac{363\,400}{230}$$
$$= 1580 \text{ mm}$$

Development of 90° conical reducing elbow (equal tapering)

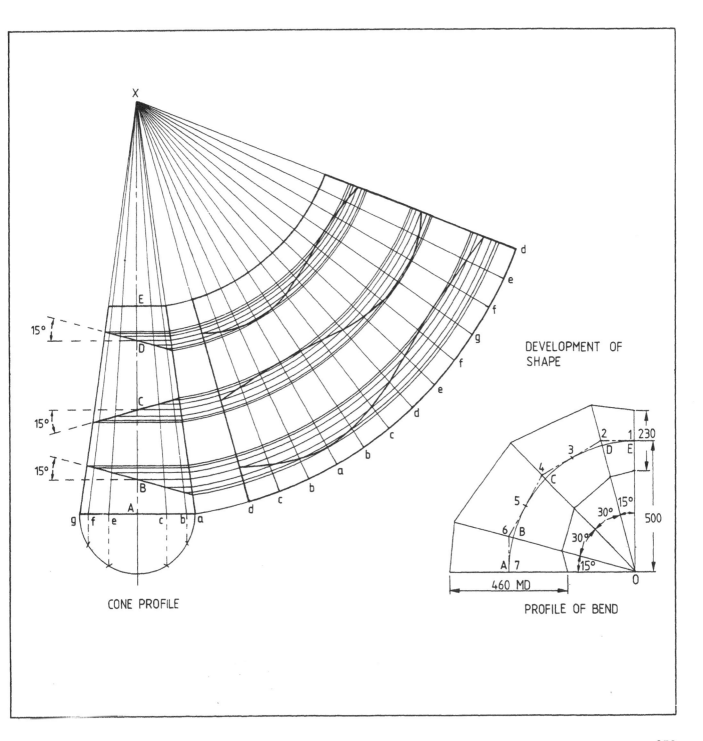

15°

15°

15°

X

E
D

C

B
A

g f e   c b' a

CONE PROFILE

d c b a b c d e f g f e d c b

DEVELOPMENT OF
SHAPE

2 1 230
3 D E
4
C
5
6 B
A 7

30° 15°
30°
15° 500

0

460 MD

PROFILE OF BEND

253

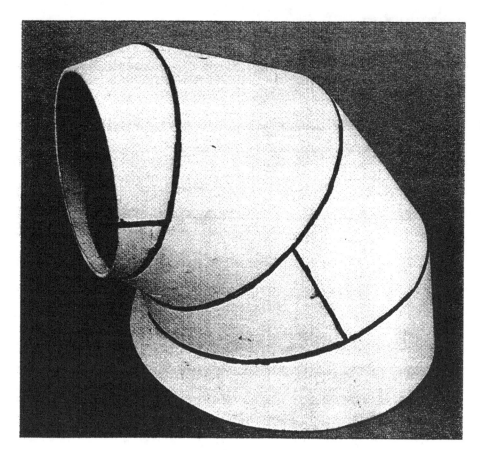

Fig. 11.2 Tapered lobsterback

Fig. 11.3 Completed job on site (circular pipe meeting conical shaped pipe meeting vertical circular pipe)

Practical example

Section of workshop drawing showing use of
frustum of right cone.

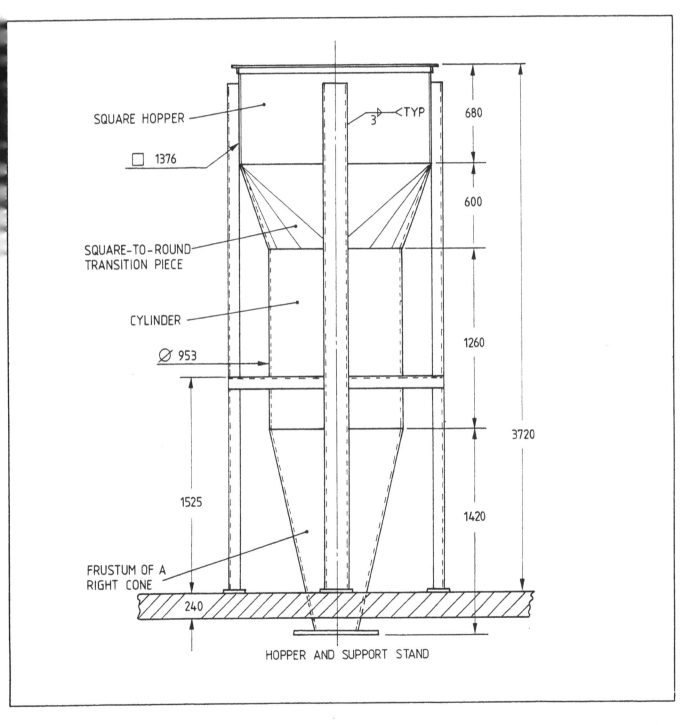

SQUARE HOPPER

□ 1376

SQUARE-TO-ROUND
TRANSITION PIECE

CYLINDER

Ø 953

FRUSTUM OF A
RIGHT CONE

240

1525

680

600

1260

3720

1420

3 < TYP

HOPPER AND SUPPORT STAND

NOTES

12

Combined radial and parallel line development

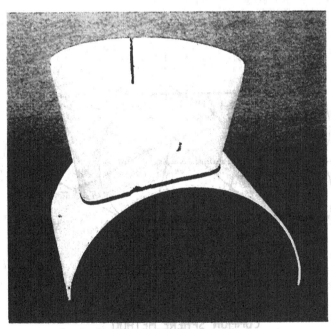

Fig. 12.1 Conical branch off centre to cylinder

In the past two chapters we have been using these two development methods by themselves, and in this chapter we combine both methods for the developments where conical and cylindrical shapes join.

This chapter will show all developments as single line drawings (no thickness shown) and you will have to apply these jobs to the particular types of fit-ups (as in Chapter 10, pp. 184–5) using the diameters required (OD, MD or ID).

This chapter will cover a variety of types of conical and cylindrical type connections which should give you enough knowledge to apply these principles to various other types of conical and cylindrical connections.

Note

Do not measure from the diagrams except in exercises, because many figures are indication diagrams only and are not drawn to scale. All measurements on the diagrams are in millimetres unless stated otherwise.

Cone and two horizontal cylinders intersecting with a common sphere

Construction

1. Draw the front and end views of the cylinder and cone showing the two views of the common sphere touching both surfaces in each view.

   Note The common sphere in the front view is also the end view of the cylinder.

2. The point of tangency indicated on the end view is horizontal to where the lines of intersection meet on the front view.

   Note It is not really necessary to draw the end view in order to find the line of intersection or the front view. The two straight lines which form the intersection may simply be drawn from one side to the other as shown and they will cross at the point of tangency.

Cone and two cylinders (one horizontal and one inclined) intersecting with a common sphere

Construction

1. Draw the front view showing the three s[*] faces *A*, *B* and *C* in their correct positio[*] each touching the common sphere. Note t[*] axes all meet at a common point *O* which [*] the centre of the common sphere.

2. Project each surface on until it intersec[*] both of the other two surfaces. These int[*] sections are shown as points *a*, *b*, *c*, *d*,[*] and *f*.

3. The two cylinders *B* and *C* alone would ha[*] intersected along *ad*, while the cylinder[*] and cone *A* alone would have intersect[*] along *be* and cylinder *C* and cone *A* along[*]

4. These lines of intersection cross at a co[*] mon point labelled *X*.

5. The portion of these lines which form t[*] line of intersection of *A*, *B* and *C* combin[*] is made up of three parts *aX*, *cX* and *eX*[*] shown outlined below.

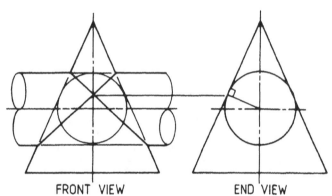

FRONT VIEW　　　END VIEW

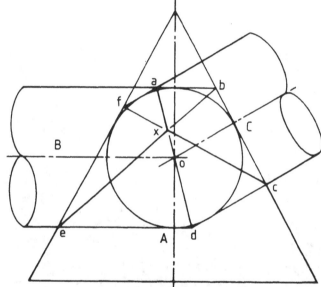

LINE OF INTERSECTION OF CYLINDER AND CONE
COMMON SPHERE METHOD

Note The cutting plane method involves drawing a series of horizontal cutting planes each of which cuts through both the intersecting surfaces, for example, a cone (to give a circle) and a cylinder (to give a rectangle).

Cone and two horizontal cylinders intersecting

Construction

1. Draw the front and end views showing the relative positions of the cone and cylinder on each view. The front view cannot be completed but the end view can.
2. Divide the end of the cylinder on the end view into twelve equal parts numbered 0, 1, 2, 3, 4, 5, 6 as shown.
3. Project these points across to the front view to represent a series of horizontal cutting planes.

4. Draw a partial half top view showing the semicircles representing the sections of the front view formed by the cutting planes. These planes are projected downwards from the front view as shown.
5. Project the points 0, 1 ... 6 from the end view down to the partial half top view to intersect the corresponding section semicircles and give the points 0, 1 ... 6 on both sides of this view.
6. Join these points with a smooth curve to outline half the line of intersection on each side of the cone.
   Note Only the line of intersection is shown on this view, that is, no cylinder is shown.
7. Project from the points 0, 1 ... 6 on the half top view up to the corresponding section plane on the front view to give points on the line of intersection. Join with a smooth curve to outline the line of intersection on both sides.

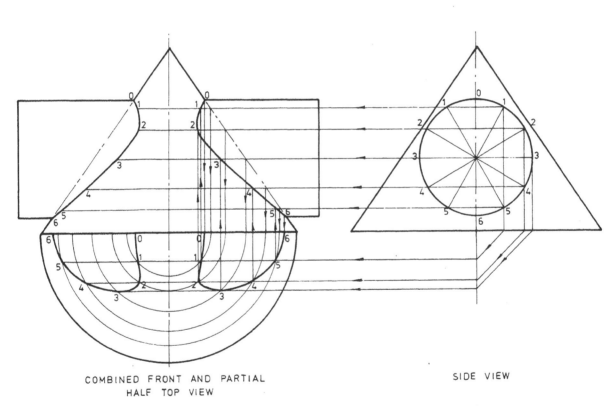

COMBINED FRONT AND PARTIAL
HALF TOP VIEW

SIDE VIEW

LINE OF INTERSECTION OF CYLINDER AND CONE
CUTTING PLANE METHOD

Vertical cylindrical branch intersecting with right cone

Operation sequence

Top view and front view
1. Draw in outline of front view.
2. Draw in top view.
3. Divide top view of cylindrical branch into twelve equal spaces and number points 1 to 7 to 1.
4. Transfer the points 2, 3, 4, 5, 6 on to front view, parallel to vertical centre line and number 1' to 7'.
5. Using A as centre and points 2, 3, 4, 5, 6 as radius, swing arcs to intersect centre line of top view at points 2^2 to 6^2.
6. Transfer points 2^2 to 6^2 from top view parallel to vertical centre line to bisect front view side line AC to locate points 2^3 to 6^3.
7. Square from centre line points 2^3 to 6^3 on side line AC to locate points 2^4 to 6^4 on vertical development lines 2' to 6'.
8. Draw in line of penetration through points 1^2, 2^4, 3^4, 4^4, 5^4, 6^4, 7^2 to complete the front view.

Development pattern of cylinder
1. Draw in rectangle, MC long by a width wider than longest developed line 7^1-7^2.
2. Divide length of rectangle into twelve equal spaces and number line 1' to 7' to 1'.
3. Transfer distance of 1^1-1^2, 2^1-2^4, 3^1-3^4, 4^1-4^4, 5^1-5^4, 6^1-6^4, and 7^1-7^2 from front view to corresponding lines of development on pattern.
4. Draw in line through points 1^2, 2^4, 3^4, 4^4, 5^4, 6^4, 7^2 to complete the true development pattern of cylinder.

Development of cone with hole developed to true shape
1. Using slope length AB as radius, point A^2 as centre, scribe an arc A^2B^2.
2. Measure around the arc the calculated MC and join to A^2 to form development cone $A^2 B^2 B^1 A^2$.
3. Mark in centre line of cone.
4. Using A^2 as centre and distances $A1^2$, $A2^3$, $A3^3$, $A4^3$, $A5^3$, $A6^3$ and $A7^2$, as radii, scribe arcs across centre line of conical pattern.
5. Measure around arcs on pattern, the curved measurements from the top view to locate points 2^5 to 6^5.

6. Draw in line through points 1^2 to 7^2 usin points 2^5 to 6^5 to complete the pattern of th cone with the development shape of th hole.

Calculations

Cylinder

$$MC = MD \times \pi$$
$$= 260 \times 3.1416$$
$$= 816.816 \text{ mm}$$
say, 817 mm

Cone

$$MC = MD \times \pi$$
$$= 750 \times 3.1416$$
$$= 2356.2 \text{ mm}$$
say, 2356 mm

Development of vertical cylindrical branch intersecting with right cone

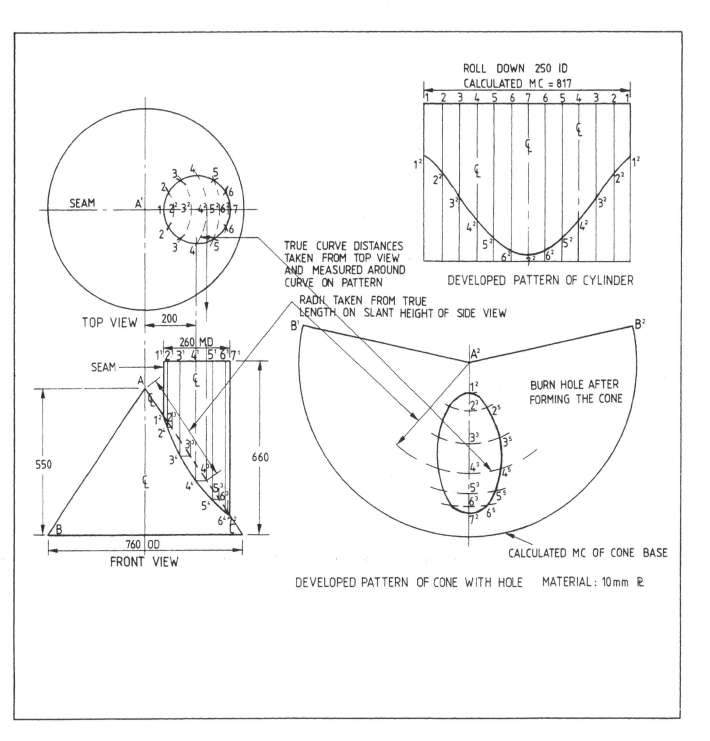

ROLL DOWN 250 ID
CALCULATED MC = 817

DEVELOPED PATTERN OF CYLINDER

TRUE CURVE DISTANCES TAKEN FROM TOP VIEW AND MEASURED AROUND CURVE ON PATTERN

RADII TAKEN FROM TRUE LENGTH ON SLANT HEIGHT OF SIDE VIEW

SEAM

TOP VIEW   200

SEAM

550

660

760 OD

FRONT VIEW   260 MD

BURN HOLE AFTER FORMING THE CONE

CALCULATED MC OF CONE BASE

DEVELOPED PATTERN OF CONE WITH HOLE   MATERIAL: 10mm R

Horizontal cylinder connecting to right cone

Operation sequence

1. Draw in front view (outline).
2. Draw in top view (outline).
3. Draw in development lines on cylinders of both views.
4. Transfer points 0^2, 1^3, 2^3, 3^3, 4^3, 5^3, 6^2 from front view up parallel to vertical centre line to locate points on horizontal centre line of top view.
5. Using point A in top view as centre, scribe all located points to locate points 0^1 to 6^1 to form true shape of opening in top view.
6. Draw in vertical parallel lines from points 0^1 to 6^1 in top view to locate points 1^2 to 5^2 in front view to locate line of penetration.

Development pattern of cylinder

1. Draw in rectangle, MC long by a width equal to longest line 4 to 4^2 in front view.
2. Divide rectangle into twelve equal spaces, mark in centre lines and number the development lines.
3. Transfer lengths of lines 1 to 6 in front view to corresponding lines of development. (Length of lines are, for example, 3 to 3^2 line of penetration.)
4. Mark in developed shape through points 0^2–6^2–0^2 to complete the developed shape.

Developed shape of cone with development of true shape of hole for cylinder

1. Using distance AB from front view as radius, scribe long curve.
2. Locate point B^1 as start point and measure calculated MC of cone base around curve to locate point B^2.
3. Scribe in curves 6^2, 5^3, 4^3, 3^3, 2^3, 1^3, 0^2 from centre point A.

4. Transfer curve distances from top view to locate points 1^1 to 5^1 on pattern.
5. Draw in line through points 0^1, 1^1, 2^1, 3^1, 4^1, 5^1, 6^1 to complete developed shape of hole in pattern for cone.

Calculations

Cylinder

$$MC = D \times \pi$$
$$= 380 \times 3.1416$$
$$= 1193.808 \text{ mm}$$
$$\text{say, } 1194 \text{ mm}$$
$$\tfrac{1}{12} C = 99.5 \text{ mm}$$

Cone

$$MC = D \times \pi$$
$$= 760 \times 3.1416$$
$$= 2387.616 \text{ mm}$$
$$\text{say, } 2388 \text{ mm}$$

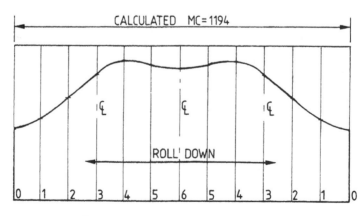

DEVELOPED SHAPE OF CYLINDER

Development of horizontal cylinder connecting to right cone

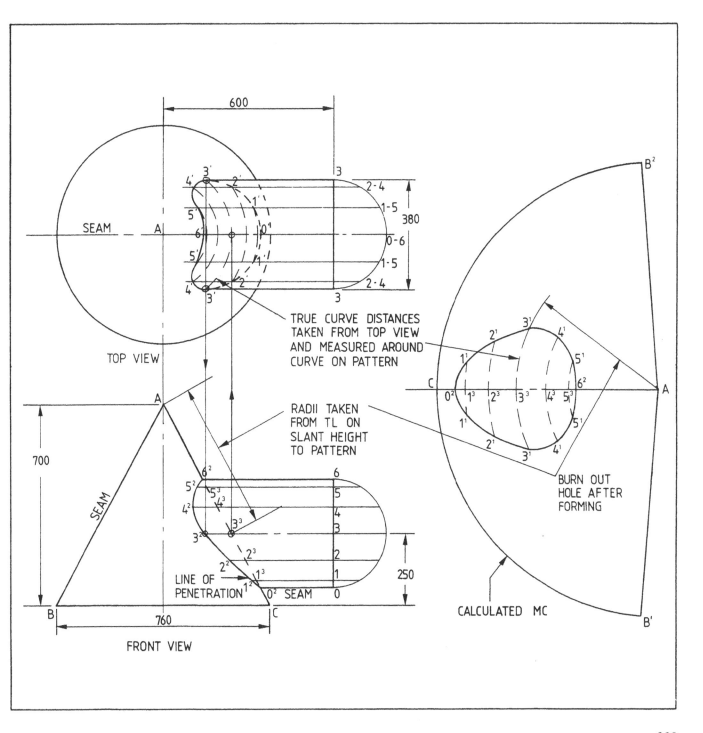

600

3

2-4

1-5

380

0-6

1-5

2-4

SEAM   A

TOP VIEW

TRUE CURVE DISTANCES
TAKEN FROM TOP VIEW
AND MEASURED AROUND
CURVE ON PATTERN

RADII TAKEN
FROM TL ON
SLANT HEIGHT
TO PATTERN

700

SEAM

6

5

4

3

2

1

250

LINE OF
PENETRATION

SEAM   0

B   760   C

FRONT VIEW

B²

C   A

BURN OUT
HOLE AFTER
FORMING

CALCULATED MC

B'

Breeching piece: Cylinder and cones around common sphere

Operation sequence

1. Lay out front view of breech piece.
2. Divide up cylinder section for parallel line development.
3. Calculate apex height of cone and position apex point A.
4. Draw in extended base section of cone and divide up for radial line development.
5. Develop true shape of cylindrical section A using parallel line development.
6. Develop true shape of conical section B and C using radial line development (same as for truncated right cone).

Calculations

Extended cone base (0–6)

C = D × π
= 550 × 3.1416
= 1727.88 mm
  say, 1728 mm

Cylindrical section

C = D × π
= 400 × 3.1416
= 1256.64 mm
  say, 1257 mm

Apex height of cone

$$H = \frac{D \times h}{D - d}$$

$$= \frac{550 \times 880}{550 - 200}$$

$$= \frac{484\,000}{350}$$

= 1382.8571 mm
  say, 1383 mm

Fig. 12.2 Conical section connecting two cylindrical sections

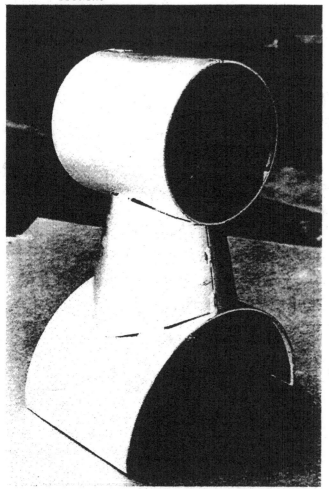

Development of breeching piece:
Cylinder and cones around common
sphere

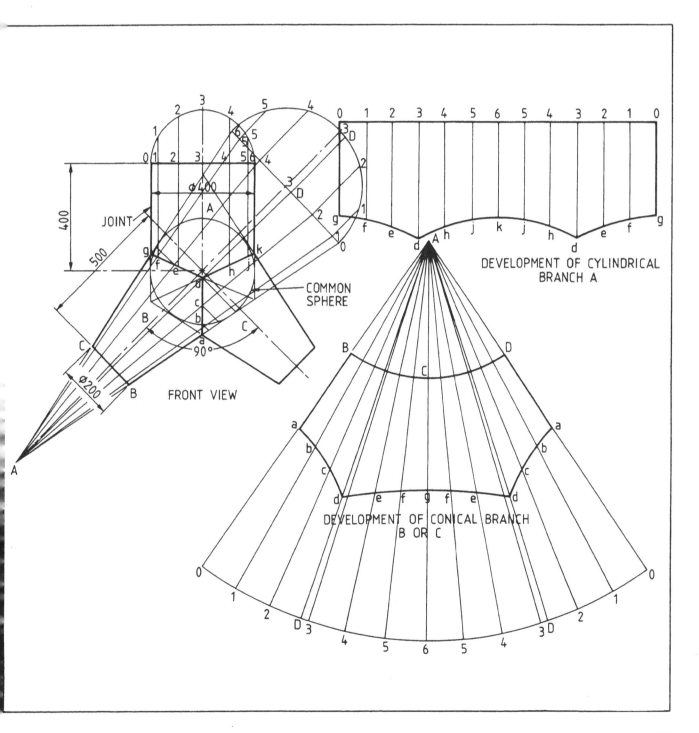

JOINT

φ400

A

COMMON
SPHERE

FRONT VIEW

90°

φ200

DEVELOPMENT OF CYLINDRICAL
BRANCH A

DEVELOPMENT OF CONICAL BRANCH
B OR C

The following pages show these single line drawings (no thickness shown):

1. Truncated right cone intersecting with a horizontal cylinder.
2. A right conical branch at 90° to the centre line.
3. A right cone joining two unequal diameter pipe lines.
4. Vertical cylinder on right cone.
5. Horizontal cylinder on right cone.

As the developments of these pages are similar to the preceding pages, we have not included any descriptions with them. Any problems encountered while developing these can be cleared up by referring back to previous pages in this chapter.

Calculations required for following pages

Apex height, $H = \dfrac{D \times h}{D - d}$

Circumference, $C = D \times \pi$

Development of truncated right cone and intersecting horizontal cylinder

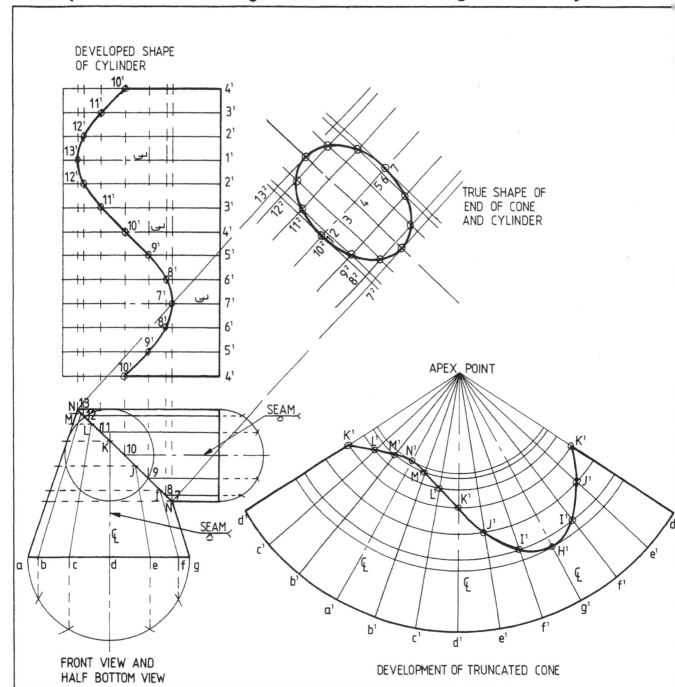

DEVELOPED SHAPE OF CYLINDER

TRUE SHAPE OF END OF CONE AND CYLINDER

APEX POINT

SEAM

SEAM

FRONT VIEW AND HALF BOTTOM VIEW

DEVELOPMENT OF TRUNCATED CONE

Development of right cone intersecting cylinder at 90° and on centre

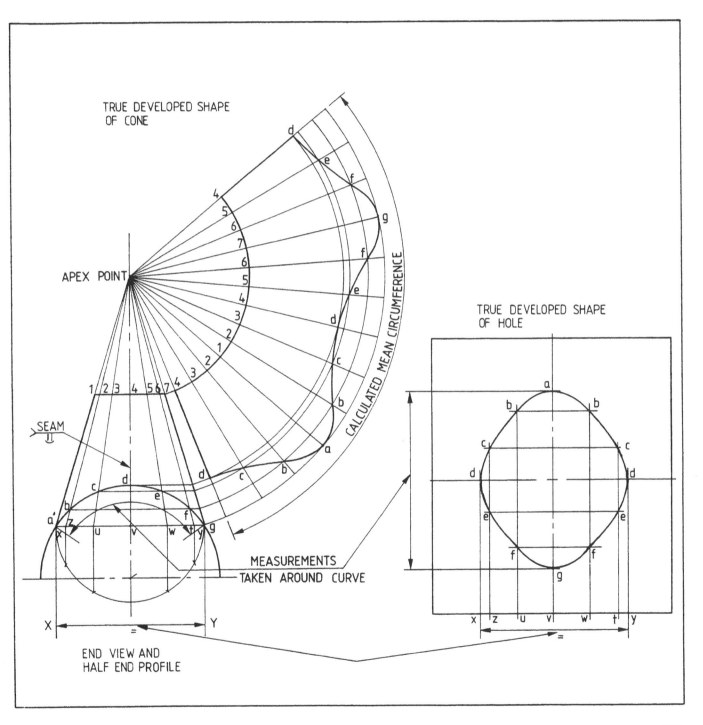

TRUE DEVELOPED SHAPE OF CONE

APEX POINT

SEAM

CALCULATED MEAN CIRCUMFERENCE

TRUE DEVELOPED SHAPE OF HOLE

MEASUREMENTS TAKEN AROUND CURVE

END VIEW AND HALF END PROFILE

Development of right cone joining two unequal diameter cylinders

For the development of holes in the cylinders refer to the method on the previous page.

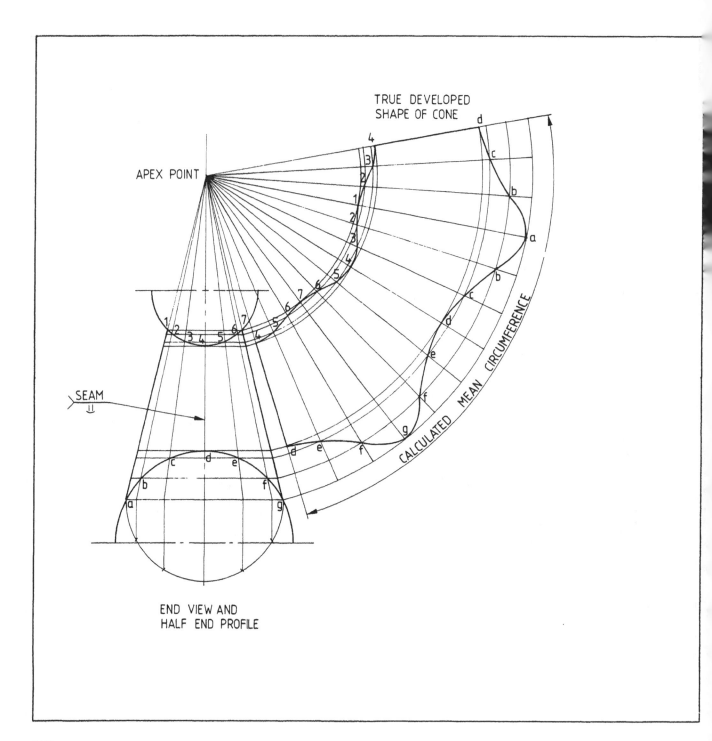

TRUE DEVELOPED SHAPE OF CONE

APEX POINT

CALCULATED MEAN CIRCUMFERENCE

SEAM

END VIEW AND HALF END PROFILE

Alternative method for development of vertical cylinder intersecting a right cone

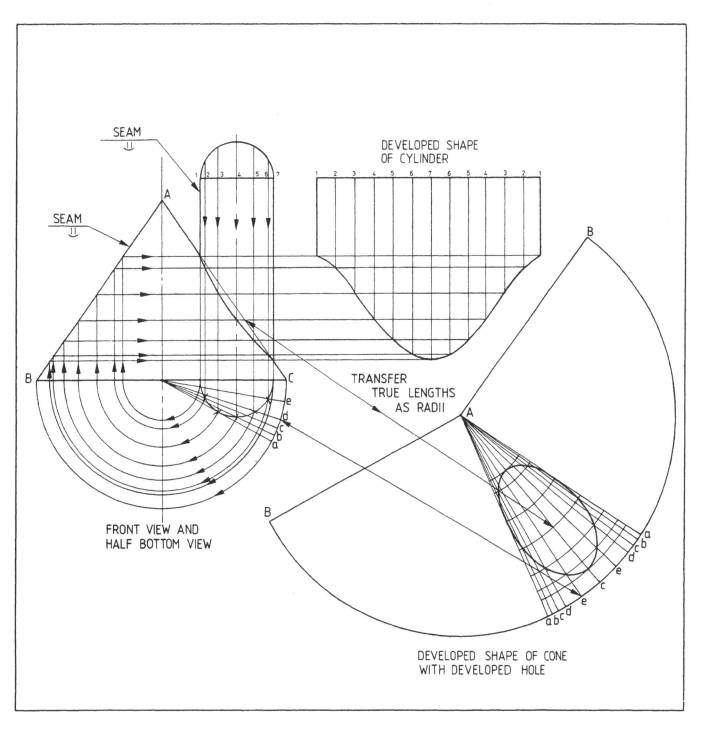

SEAM

SEAM

DEVELOPED SHAPE OF CYLINDER

FRONT VIEW AND HALF BOTTOM VIEW

TRANSFER TRUE LENGTHS AS RADII

DEVELOPED SHAPE OF CONE WITH DEVELOPED HOLE

Alternative method for development of horizontal cylinder intersecting a right cone

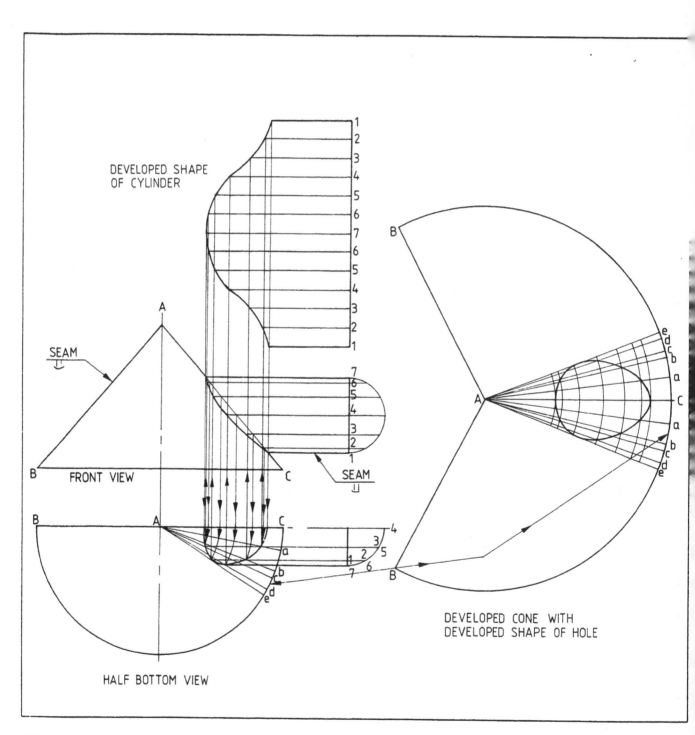

DEVELOPED SHAPE OF CYLINDER

SEAM

FRONT VIEW

A

B

C

SEAM

HALF BOTTOM VIEW

DEVELOPED CONE WITH DEVELOPED SHAPE OF HOLE

Fig. 12.3 Setting up and checking vertical and horizontal branches on a right cone

Fig. 12.4 Circular tankwork showing associated pipework in background

Exercises for developing connecting cones and cylinders

1. Complete the following exercise showing:
   (a) line of penetration;
   (b) development of conical section with hole;
   (c) developed shape of cylindrical branch;
   (d) calculation of MC of cylindrical branch;
   (e) calculation of MC of base of cone.

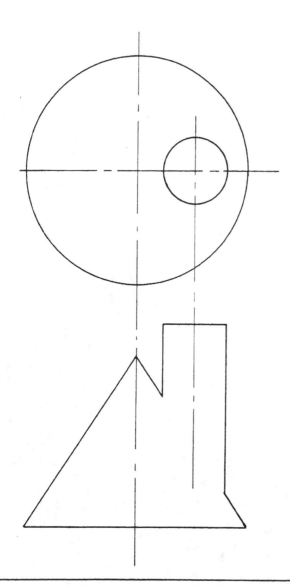

Exercises (cont'd)

2. Complete the following exercise showing:
   (a) line of penetration;
   (b) development of conical section with hole;
   (c) developed shape of cylinder;
   (d) calculation of MC of cylindrical branch;
   (e) calculation of MC of base of cone.

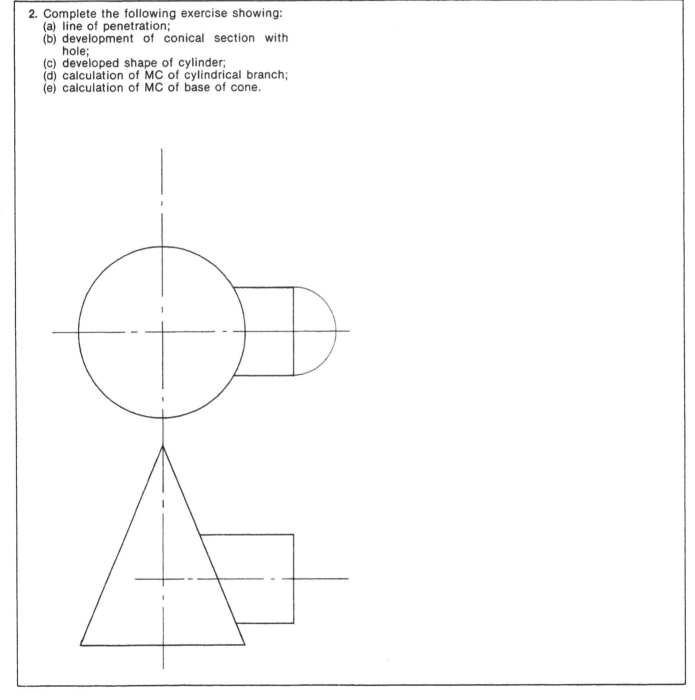

Exercises (cont'd)

3. Complete the following exercise showing:
  (a) development of cylindrical section *A*;
  (b) development of conical section *B* or *C*;
  (c) calculation of MC of cylindrical section;
  (d) calculation of apex height of conical section;
  (e) calculation of MC of base of cone;
  (f) calculation of MC of top of cone.

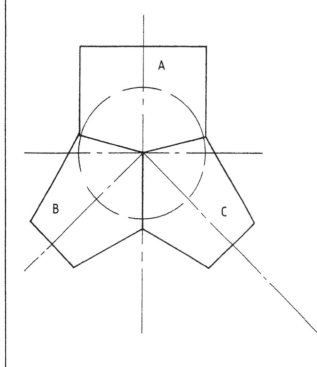

Exercises (cont'd)

4. Complete the following exercise showing:
- (a) development of cylindrical section;
- (b) development of conical section;
- (c) development of true shape of opening;
- (d) calculation of MC of cylinder;
- (e) calculation of apex height of conical section;
- (f) calculation of MC of base of cone.

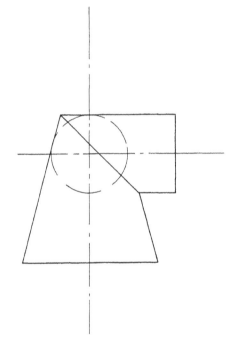

Exercises (cont'd)

5. Complete the following exercise showing:
(a) developed shape of conical section;
(b) calculation of MC of base of conical section;
(c) calculation of MC of top of conical section;
(d) calculation of apex height of conical section;
(e) developed shape of hole in small cylinder;
(f) developed shape of hole in large cylinder.

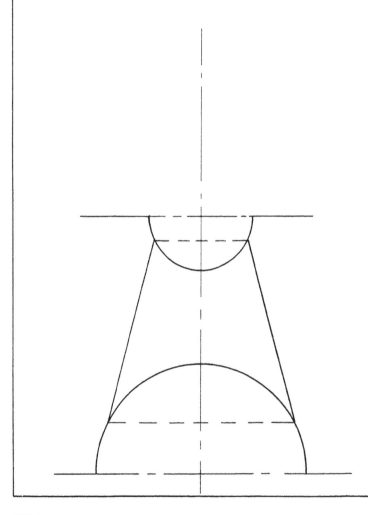

13

Triangulation of transition pieces

Triangulation *is a form of development used when the shape to be developed cannot be developed using parallel line development or radial line development. The principle of triangulation is:*

To develop a pattern by division of the surface of the component into a number of triangles, determine the true size and shape of each, and then lay them down side by side in the correct order to produce a pattern.

The golden rule of triangulation is:

Place the top or bottom view length of a line at right angles to its vertical height. The diagonal will represent its true length.

In this chapter we will start with a step-by-step procedure for a few basic-shaped patterns. Then we will move on to complete developments of these basic shapes and other shaped transition pieces.

As there are many applications of shapes for triangulation, we will only be giving you the developments for some of these. If you understand the principle of triangulation and you can apply it, then there is no problem with the development of the other shapes.

Note
Do not measure from the diagrams except in exercises, because many figures are indication diagrams only and are not drawn to scale. All measurements on the diagrams are in millimetres unless stated otherwise.

Fig. 13.1 Transition piece: square to round

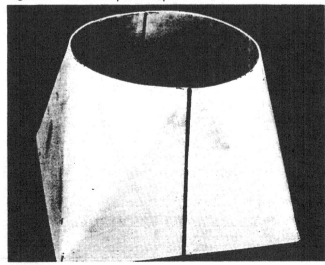

Part A
Triangulation of transition pieces with the top and base planes parallel

Problem

To develop half pattern for concentric square to round transition piece by triangulation.

Procedure

1. Lay out half top view using MD for circular end and inside measurements for square end.
2. Divide half circle into six equal parts.
3. Construct true length diagram using the vertical height of the transition piece and distances from half top view.

Calculation

$$MC = MD \times \pi$$
$$= 30 \times 3.1416$$
$$= 94.248 \text{ mm}$$
$$\tfrac{1}{2} MC = 47.124 \text{ mm}$$
$$\tfrac{1}{4} MC = 23.562 \text{ mm}$$
$$\tfrac{1}{12} MC = 7.854 \text{ mm}$$

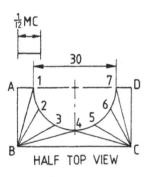

HALF TOP VIEW

TRUE LENGTH DIAGRAM

VERTICAL HEIGHT

X   D7 B2-3 B1-4
    A1 C5-6 C4-7

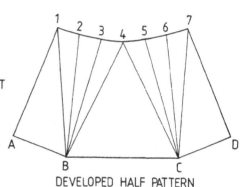

DEVELOPED HALF PATTERN

1

(a) Draw in base line equal to length *BC* in half top view.
(b) Use distances *B4* and *C4* from half top view and place on true length diagram from point *X*.
(c) Using true length distances *ZB4* and *ZC4*, and points *B* and *C* as centres, scribe arcs to intersect at point 4.

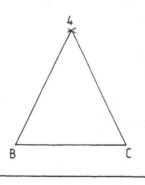

2

(a) Using ¹⁄₁₂ of MC, scribe arcs both sides of point 4.
(b) Use distances *B3* and *C5* from half top view and place on TL diagram from point *X*.
(c) Using true length distances *ZB3* and *ZC5* and points *B* and *C* as centres, scribe arcs to locate points 3 and 5.

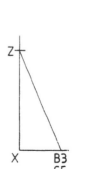
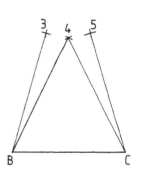

3

(a) Scribe ¹⁄₁₂ MC from points 3 and 5.
(b) Use distances *B2* and *C6* from half top view and place on TL diagram from point *X*.
(c) Using true length distances *ZB2* and *ZC6* and points *B* and *C* as centres, scribe arcs to locate points 2 and 6.
(d) Check around curve from points 2 to 5 to be equal to one-quarter of MC.

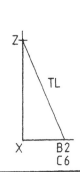

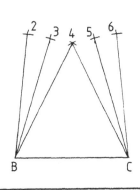

4

(a) Scribe ¹⁄₁₂ MC from points 2 and 6.
(b) Use distances *B1* and *C7* from half top view and place on TL diagram from point *X*.
(c) Using true length distances *ZB1* and *ZC7* and points *B* and *C* as centres, scribe arcs to locate points 1 and 7.
(d) Draw in curved line from points 1 to 7 with a flexible straightedge.

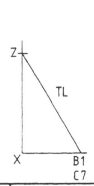

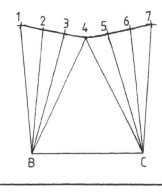

5

(a) Using distances *AB* and *CD* from half top view and points *B* and *C* as centres, scribe arcs.
(b) Using distances *A1* and *D7* from half top view, place on TL diagram from point *X*.
(c) Using true length distances *ZA1* and *ZD7* and points 1 and 7 as centres, scribe arcs to locate points *A* and *D*.

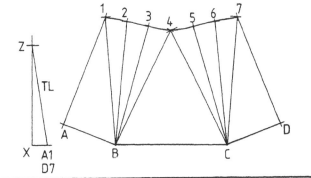

6

Checking techniques
(a) Measure diagonals; 1*D* and 7*A* should be equal.
(b) Measure around curve points; 1 to 7 should equal half mean circumference.
(c) Check that angles 1*AB* and 7*DC* are 90° by angle in a semicircle method:
  (i) Bisect lines 1*B* and 7*C* to locate centre points *O*.
  (ii) Using point *O* as centre and *O1*, *OB*, *O7* and *OC* as radii, scribe semicircle and points *A* and *D* should intersect on semicircle to prove they are 90°.

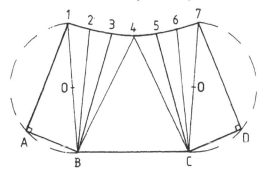

Problem

To develop half pattern for an offset square to round transition piece by triangulation.

Procedure

1. Lay out half top view using MD for circul[ar] end and inside measurement for square en[d].
2. Divide semicircle of half top view into s[ix] equal parts and join points to corners [of] square section.
3. Construct the true length diagram using th[e] vertical height of the transition piece plu[s] the given lengths from the half top view.

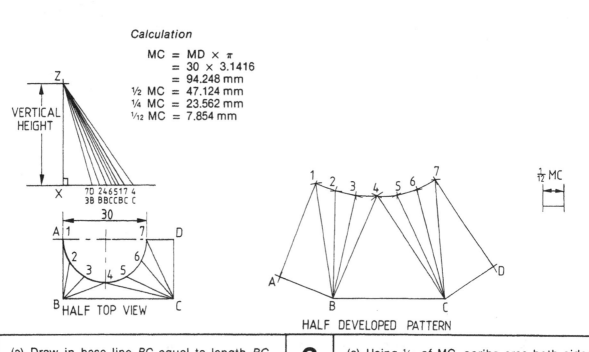

Calculation

$$MC = MD \times \pi$$
$$= 30 \times 3.1416$$
$$= 94.248 \text{ mm}$$
$$\tfrac{1}{2} MC = 47.124 \text{ mm}$$
$$\tfrac{1}{4} MC = 23.562 \text{ mm}$$
$$\tfrac{1}{12} MC = 7.854 \text{ mm}$$

VERTICAL HEIGHT

30

HALF TOP VIEW

HALF DEVELOPED PATTERN

$\frac{1}{12}$ MC

1

(a) Draw in base line *BC* equal to length *BC* from half top view.
(b) Using distances B4 and C4 from half top view, place on true length diagram from point *X*.
(c) Using true length distance ZB4 and point *B* as centre, scribe radius.
(d) Using true length distance ZC4 and point *C* as centre, scribe radius to cut first radius to locate point 4.

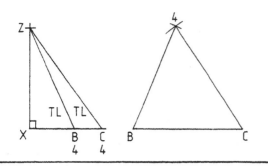

2

(a) Using ¹⁄₁₂ of MC, scribe arcs both sides of point *B*.
(b) Using distances B3 and C5 from half top view, place on true length diagram from point *X*.
(c) Using true length distance ZB3 and point *B* as centre, scribe arc to locate point 3.
(d) Using true length distance ZC5 and point *C* as centre, scribe arc to locate point 5.

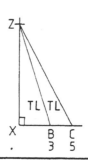
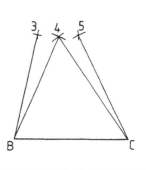

280

3

(a) Using ¹⁄₁₂ MC as radius, scribe arcs from points 3 and 5.
(b) Using distances *B2* and *C6* from half top view place on true length diagram from point *X*.
(c) Using true length distance *ZB2* and point *B* as centre, scribe arc to locate point 2.
(d) Using true length distance *ZC2* and point *C* as centre, scribe arc to locate point 6.
(e) Check points 2 to 5 and 6 to 3 around curve to be equal to one-quarter calculated mean circumference.

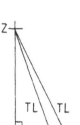
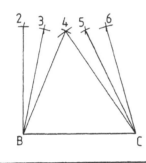

4

(a) Using ¹⁄₁₂ MC as radius, scribe arc from points 2 and 6.
(b) Using distances *B1* and *C7* from half top view, place on true length diagram from point *X*.
(c) Using true length distance *ZB1* and point *B* as centre, scribe arc to locate point 1.
(d) Using true length distance *ZC7* and point *C* as centre, scribe arc to locate point 7.

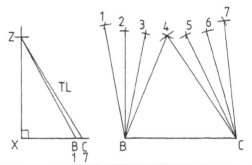

5

(a) Using distances *AB* and *CD* from half top view, scribe arcs from points *B* and *C*.
(b) Use distance *7D* from half top view and place it on true length diagram from point *X*.
(c) Using true length *Z7D* and point 7 as centre, scribe arc to locate point *D*.
(d) Using vertical height *ZX* as radius and point 1 as centre, scribe arc to locate point *A*.

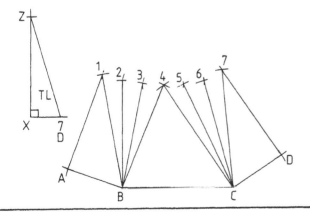

6

Join points 1 to 7 with a curved line using a flexible straightedge to complete the true developed pattern.

Checking technique
(a) Check around curved line from points 1 to 7 to equal half calculated mean circumference.
(b) Check corner *7DC* to equal 90°.
(c) Check corner *1AB* to equal 90°.

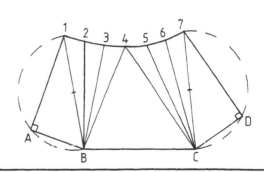

Development of concentric square to round transition piece by triangulation

Notes

1. Only need to use quarter top view, vertical height and true lengths.
2. Use inside measurements for square end.
3. Use mean dimensions for circular end.
4. Develop same as in earlier exercise.

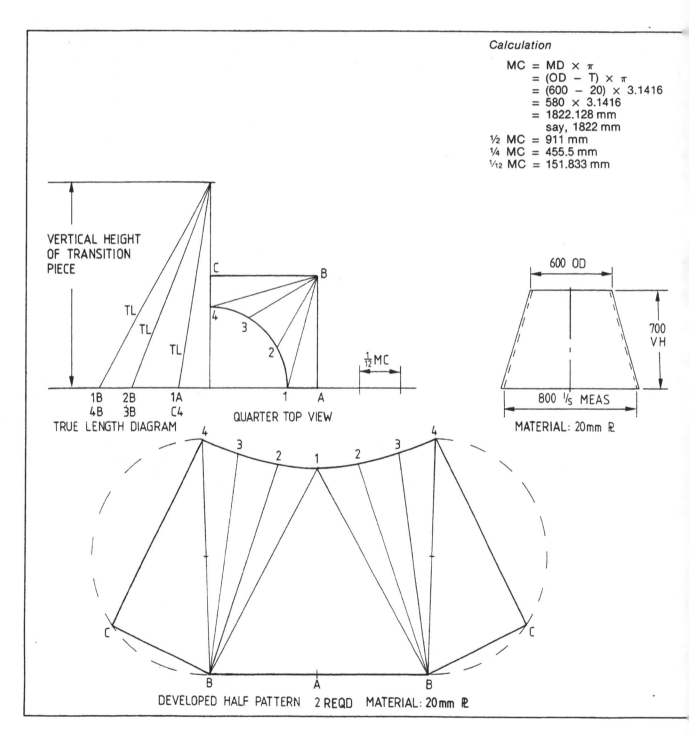

Calculation

$$MC = MD \times \pi$$
$$= (OD - T) \times \pi$$
$$= (600 - 20) \times 3.1416$$
$$= 580 \times 3.1416$$
$$= 1822.128 \text{ mm}$$
$$\text{say, } 1822 \text{ mm}$$

½ MC = 911 mm
¼ MC = 455.5 mm
1/12 MC = 151.833 mm

VERTICAL HEIGHT OF TRANSITION PIECE

TL

TL

TL

1B 2B 1A
4B 3B C4

TRUE LENGTH DIAGRAM

1/12 MC

QUARTER TOP VIEW

600 OD

700 VH

800 I/s MEAS

MATERIAL: 20mm ₽

DEVELOPED HALF PATTERN 2 REQD MATERIAL: 20mm ₽

Development of offset square to round transition piece by triangulation

Notes

1. Only need to use half top view, vertical height and true lengths.
2. Use inside dimensions for square end.
3. Use mean dimensions for circular end.
4. Develop the same sequence as the previous exercises.

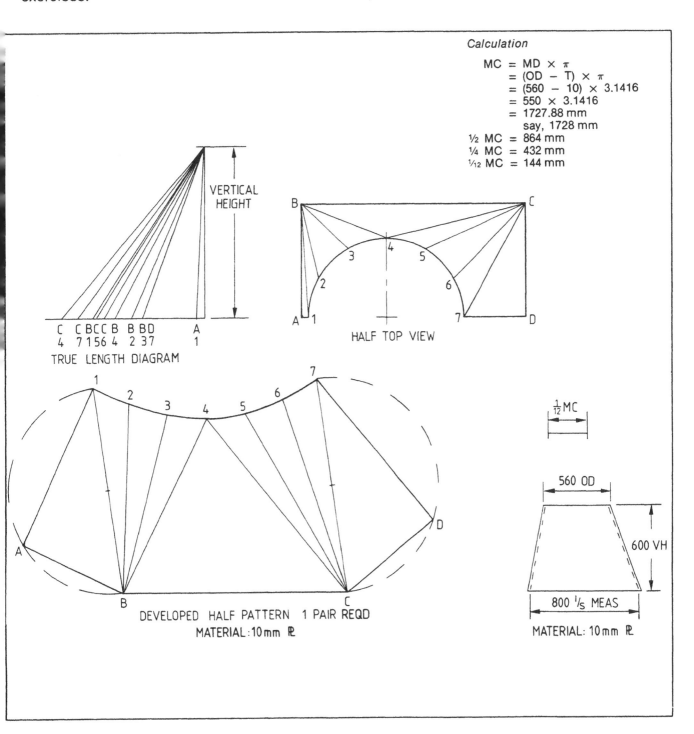

Calculation

$$MC = MD \times \pi$$
$$= (OD - T) \times \pi$$
$$= (560 - 10) \times 3.1416$$
$$= 550 \times 3.1416$$
$$= 1727.88 \text{ mm}$$
$$\text{say, } 1728 \text{ mm}$$
$$\tfrac{1}{2} MC = 864 \text{ mm}$$
$$\tfrac{1}{4} MC = 432 \text{ mm}$$
$$\tfrac{1}{12} MC = 144 \text{ mm}$$

VERTICAL HEIGHT

TRUE LENGTH DIAGRAM

HALF TOP VIEW

DEVELOPED HALF PATTERN 1 PAIR REQD
MATERIAL: 10 mm ℞

$\tfrac{1}{12}$ MC

560 OD

600 VH

800 $^1/_S$ MEAS

MATERIAL: 10 mm ℞

Development of offset rectangle to round transition piece by triangulation

Notes for layout

1. Rectangular end to inside dimensions.
2. Round end to mean dimensions.

Check

1. Curved end 1–4–7 to equal the calculated mean circumference.
2. End angles 7DC and 1AB to be 90° (angle in semicircle method).

Fig. 13.2 Transition piece: offset rectangle to roun

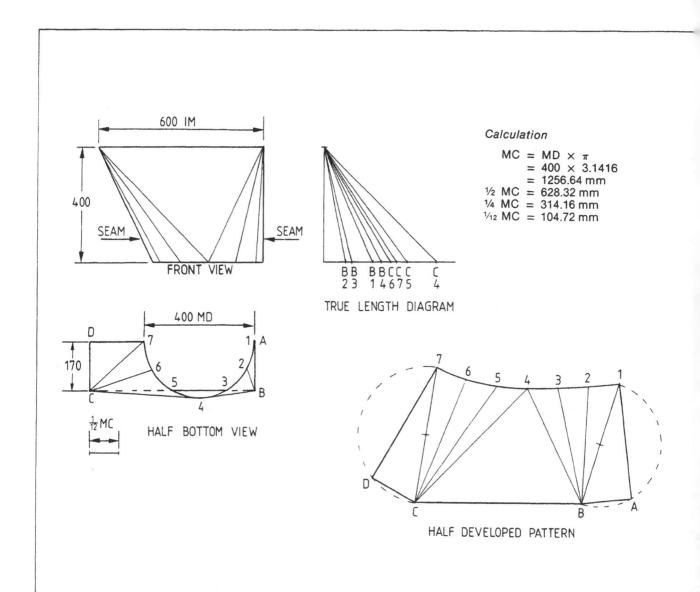

600 IM

400

SEAM   SEAM

FRONT VIEW

TRUE LENGTH DIAGRAM

B B B B C C C C
2 3 1 4 6 7 5 4

Calculation

$$MC = MD \times \pi$$
$$= 400 \times 3.1416$$
$$= 1256.64 \text{ mm}$$
$$\tfrac{1}{2} MC = 628.32 \text{ mm}$$
$$\tfrac{1}{4} MC = 314.16 \text{ mm}$$
$$\tfrac{1}{12} MC = 104.72 \text{ mm}$$

400 MD

170

$\tfrac{1}{12}$ MC

HALF BOTTOM VIEW

HALF DEVELOPED PATTERN

284

Variation of triangulation development of offset rectangle to round transition piece

Notes

1. Use MD for top.
2. Use inside measurements for base.
3. Check curve 1–7–1 is equal to calculated MC of top.

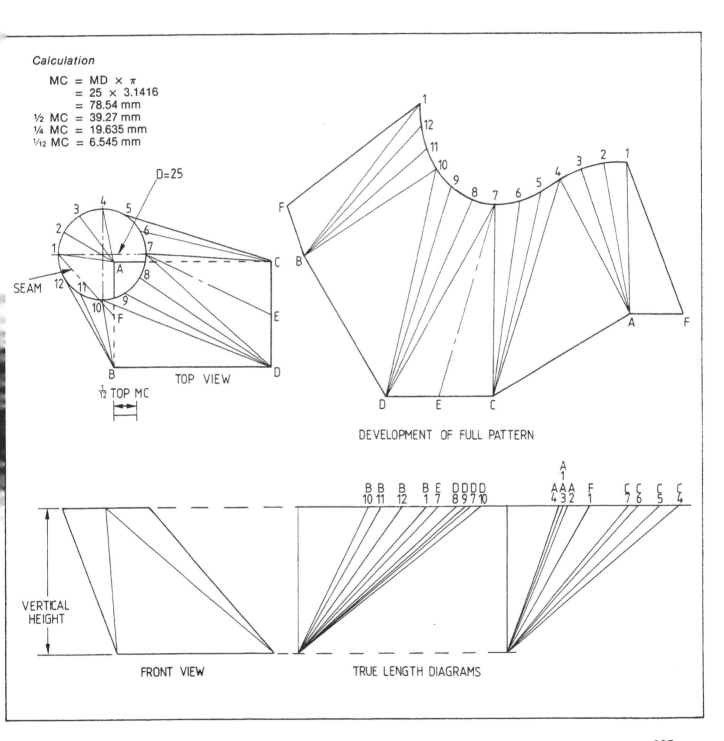

Calculation

$$MC = MD \times \pi$$
$$= 25 \times 3.1416$$
$$= 78.54 \text{ mm}$$
$$\tfrac{1}{2} MC = 39.27 \text{ mm}$$
$$\tfrac{1}{4} MC = 19.635 \text{ mm}$$
$$\tfrac{1}{12} MC = 6.545 \text{ mm}$$

D=25

SEAM

$\frac{1}{12}$ TOP MC

TOP VIEW

DEVELOPMENT OF FULL PATTERN

VERTICAL HEIGHT

FRONT VIEW

TRUE LENGTH DIAGRAMS

Development of slow tapering frustum of right cone by triangulation

Quarter top view and true length diagram (only two true lengths required).

Notes

1. Front view of job not required for development.
2. Develop using mean diameters and mean circumferences.

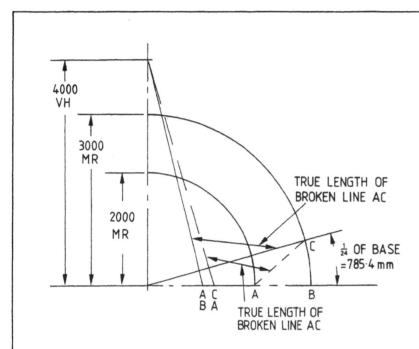

QUARTER OF TOP VIEW AND TRUE LENGTH DIAGRAM

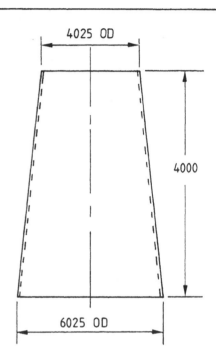

JOB DETAILS 25mm MS ℞

Calculations

MC of base = MD × π
            = (OD − T) × π
            = (6025 − 25) × 3.1416
            = 6000 × 3.1416
            = 18 849.5 mm
    ½ MC = 9424.75 mm
    ¼ MC = 4712.4 mm
    ⅛ MC = 2356.2 mm
    ¹⁄₂₄ MC = 785.4 mm

MC of top = MD × π
           = (OD − T) × π
           = (4025 − 25) × 3.1416
           = 4000 × 3.1416
           = 12 566.4 mm
    ½ MC = 6283.2 mm
    ¼ MC = 3141.6 mm
    ⅛ MC = 1570.8 mm
    ¹⁄₂₄ MC = 523.6 mm

Development of slow tapering frustum of cone by triangulation (cont'd)

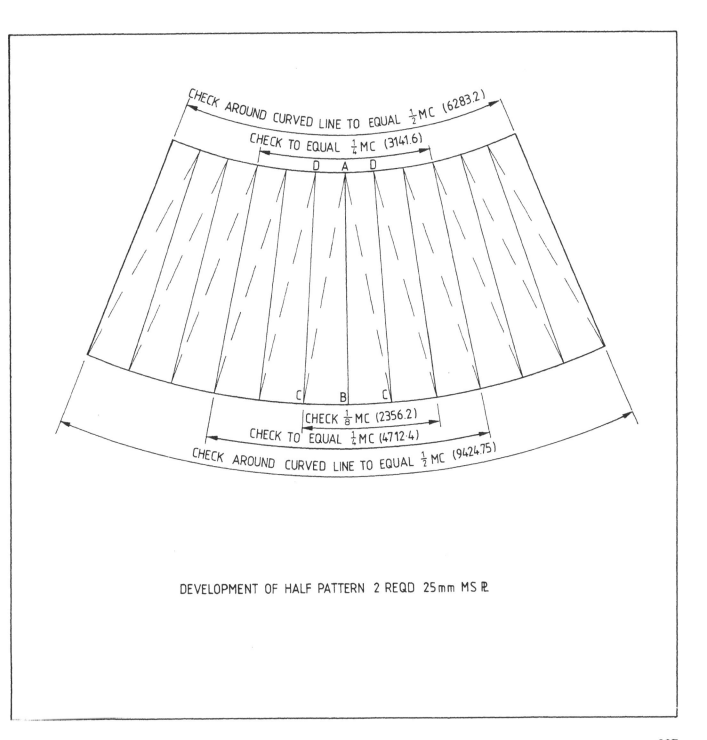

CHECK AROUND CURVED LINE TO EQUAL ½ MC (6283.2)

CHECK TO EQUAL ¼ MC (3141.6)

D   A   D

C   B   C

CHECK ⅛ MC (2356.2)

CHECK TO EQUAL ¼ MC (4712.4)

CHECK AROUND CURVED LINE TO EQUAL ½ MC (9424.75)

DEVELOPMENT OF HALF PATTERN 2 REQD 25mm MS ℞

Offset round to round by triangulation

Operation sequence

Half top view and true length diagram

1. Draw in half top view to mean dimensions.
2. Divide half top view into required number of triangles.
3. Identify points of half top view, 1 to 7 on large semicircle and *a* to *g* on small semicircle.
4. Draw up true length diagram using the vertical height of the frustum and the given lengths from the half top view.

Remember the golden rule of triangulation:

Take any given length from the top or bottom views and place it at 90° to the vertical height of the section and the diagonal or slope length is a true length.

Calculations

MC of base $= MD \times \pi$
$= (OD - T) \times \pi$
$= (1250 - 20) \times 3.1416$
$= 1230 \times 3.1416$
$= 3864.168$ mm
½ MC $= 1932.084$ mm
¼ MC $= 966.042$ mm
$\frac{1}{12}$ MC $= 322.014$ mm

MC of top $= MD \times \pi$
$= (OD - T) \times \pi$
$= (720 - 20) \times 3.1416$
$= 700 \times 3.1416$
$= 2199.12$ mm
½ MC $= 1099.56$ mm
¼ MC $= 549.78$ mm
$\frac{1}{12}$ MC $= 183.26$ mm

Pattern development

1. Draw in full line using true length *g7*.
2. Using point 7 as centre and $\frac{1}{12}$ large MC as radius, scribe arcs both sides of point 7.
3. Using point *g* as centre and true length *6g* as radius, scribe arcs to locate point 6 (draw in dotted lines to connect point *g* to 6).
4. Using point *g* as centre and $\frac{1}{12}$ small MC as radius, scribe arcs both sides of point *g*.
5. Using point 6 as centre and true length *f6* as radius, scribe arcs to locate point *f* (draw in full lines to connect point 6 to *f*).

This now completes the two triangles 6*g* and *g6f* each side of the centre line *g7*.

The rest of the pattern is developed using the same procedure and the different true lengths to complete the full pattern.

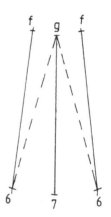

Development of offset round to round by triangulation

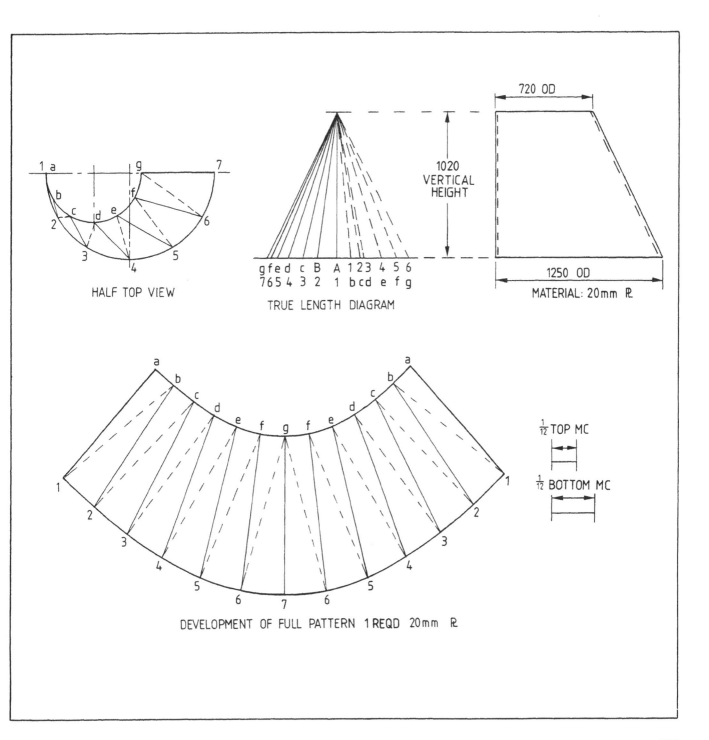

HALF TOP VIEW

TRUE LENGTH DIAGRAM

1020 VERTICAL HEIGHT

720 OD

1250 OD

MATERIAL: 20mm ℞

gfed c B A 1 2 3 4 5 6
765 4 3 2 1 bcd e f g

$\frac{1}{12}$ TOP MC

$\frac{1}{12}$ BOTTOM MC

DEVELOPMENT OF FULL PATTERN 1 REQD 20mm ℞

Rectangular base with rounded corners and circular top transition piece by triangulation

Operation sequence

Top view and true length diagram

1. As top and base are parallel, there is no need to draw a front view.
2. As the four quarters of the top view are the same, then only one quarter of the top view needs to be drawn.
3. Draw all views to mean dimensions.
4. True lengths are obtained by using the vertical height of the transition piece, given distances from top view placed at 90° to vertical height, and the diagonal becomes the true length.

Development

1. Draw in top view.
2. Divide one-quarter of top view of circular section into any number of equal sections and number the points.
3. Divide the rounded corner of the rectangles into the same amount of sections and number the points.
4. Draw up true length diagram.
5. Construct developed half pattern, starting from centre line 1A.

6. Construct triangles 1-2-A using true length
7. Continue constructing triangles as in rour to round triangulation until points 6 and are reached.
8. Construct end triangles E76 similar to er sections on square to round pattern.

Checking

1. Check around curve EAE to equal half MC top.
2. Check around curves 6–2 and 2–6 to equ mean length of base corner.
3. Check outside angle E76 to equal 90° (us angle in semicircle method).
4. Check diagonals E7 are equal.

Calculations

MC of base corners = (MR × 2) × π
= (360 × 2) × π
= 720 × 3.1416
= 2261.952 mm
½ MC = 1130.976 mm
¼ MC = 565.488 mm
1/16 MC = 141.372 mm

MC of top = MD × π
= 1020 × 3.1416
= 3204.432 mm
½ MC = 1602.216 mm
¼ MC = 801.108 mm
1/16 MC = 200.277 mm

Fig. 13.3 Transition piece: square base with rounded corners to circular top

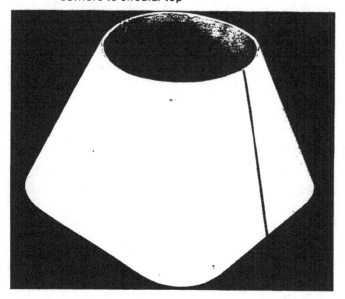

Development of rectangular base with rounded corners and circular top transition piece by triangulation

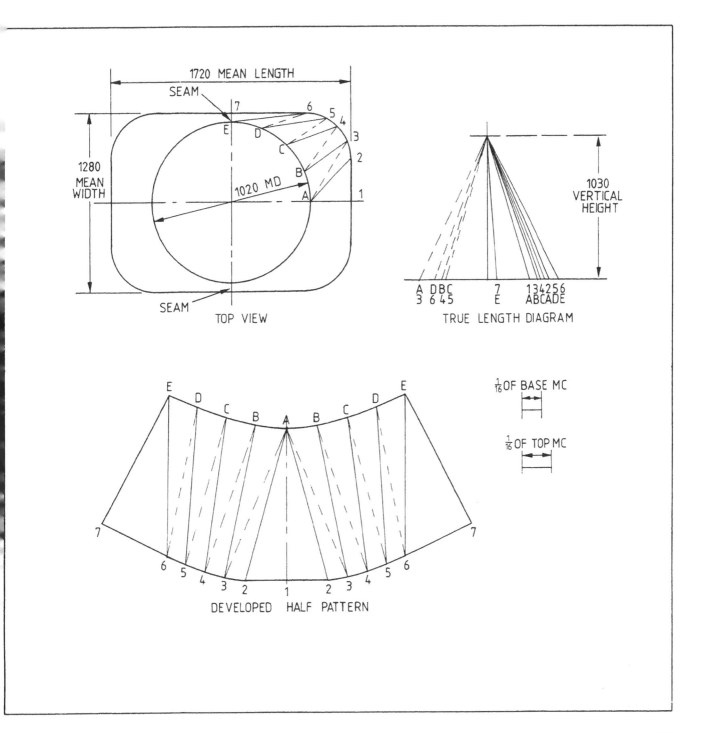

TOP VIEW

1720 MEAN LENGTH

SEAM

1280 MEAN WIDTH

1020 MD

SEAM

TRUE LENGTH DIAGRAM

1030 VERTICAL HEIGHT

A DBC   7   134256
3 6 45   E   ABCADE

$\frac{1}{16}$ OF BASE MC

$\frac{1}{16}$ OF TOP MC

DEVELOPED HALF PATTERN

Development of transition piece with square base with rounded corners and circular top by triangulation

Notes

1. Lay out to mean dimensions.
2. Check curve *AEA* to equal half calculated MC of top.
3. Check curves 2–6 and 6–2 to equal quarter MC of corner.
4. Check outside corners to equal 90° (using angle in a semicircle method).
5. Check diagonals *A*–1 are equal.

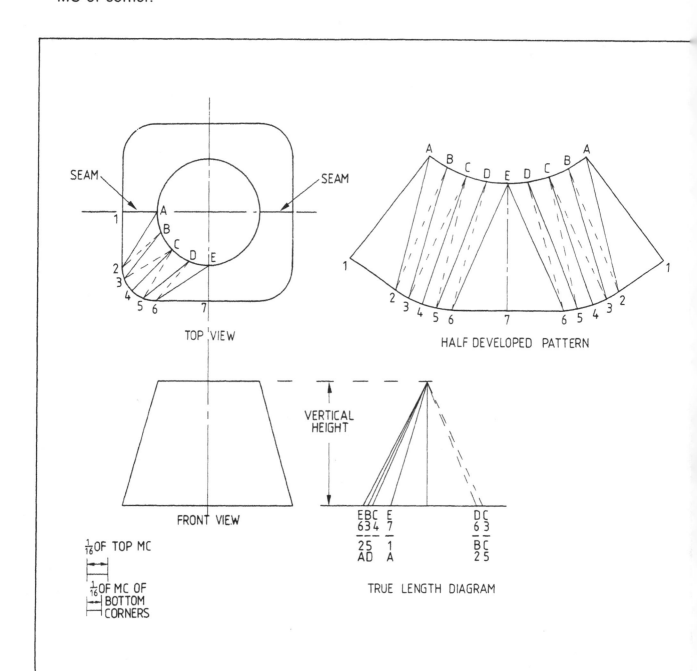

TOP VIEW

HALF DEVELOPED PATTERN

FRONT VIEW

VERTICAL HEIGHT

$\frac{1}{16}$ OF TOP MC

$\frac{1}{16}$ OF MC OF BOTTOM CORNERS

TRUE LENGTH DIAGRAM

Development of transition piece with rectangular base and obround top by triangulation

Notes

1. Use MD of top.
2. Use inside measurements for base.
3. Check top curves 1–4 and 5–8 to equal quarter of top MC.
4. Check outside corners 1AB and 8CD are 90°.
5. Check diagonal 1D is equal to 8A.

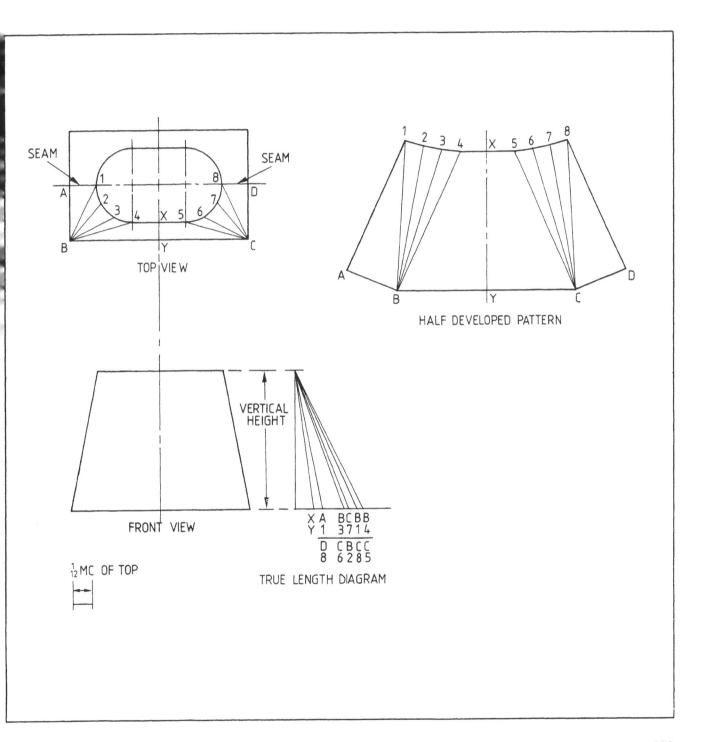

TOP VIEW

HALF DEVELOPED PATTERN

VERTICAL HEIGHT

FRONT VIEW

$\frac{1}{12}$ MC OF TOP

X A	BCBB
Y 1	3714
D	CBCC
8	6285

TRUE LENGTH DIAGRAM

Development of triangular base with rounded corners and circular top by triangulation

Notes

1. Lay out to mean dimensions.
2. Check top curve to be equal to calculated top MC.
3. Check base curves 1_1–3_1–1_1 are equal to calculated length of base curve from top view.
4. Check outside corners $EX1_1$ are 90° (angle in semicircle method).
5. Check diagonals are equal.
6. a and b represent $1/12$ of their respective cumferences of the top and bottom sectio

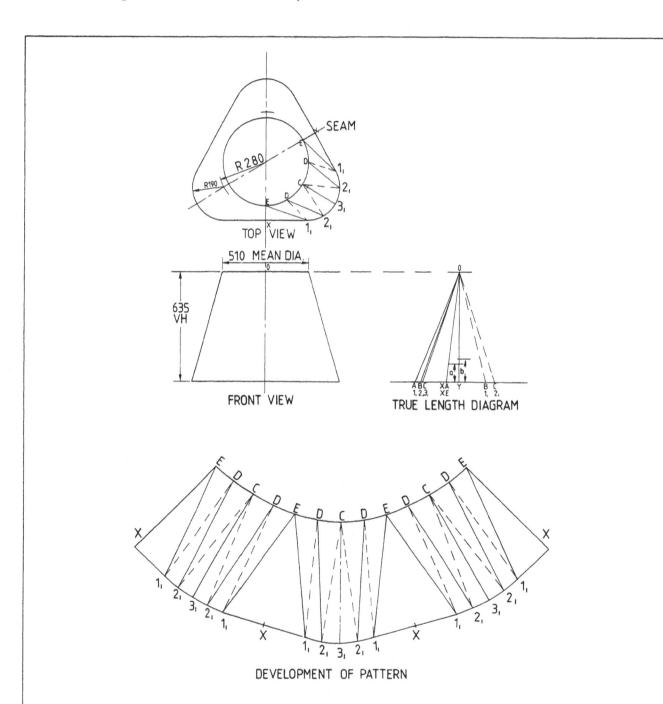

TOP VIEW

FRONT VIEW

TRUE LENGTH DIAGRAM

DEVELOPMENT OF PATTERN

Development of transition piece ellipse to round by triangulation

Operation sequence

1. On marking out plate, draw top view and front view of job as shown, using circular arcs method for marking ellipse.
2. In top view, divide one quarter section of the elliptical base and circular top into any number of equal parts and number as shown.
3. For true length diagram, erect a line equal in length to the height of the job and at right angles to a horizontal line *CD*.
4. With *A* as centre, on the true length diagram mark out the plain and dotted length obtained from the top view as shown, and number.
5. Join plain and dotted lines as shown.
6. To mark the pattern, erect a vertical centre line on the plate equal in length to the line 1–1₁ obtained from the true length diagram.
7. Mark out on each side of the centre line 1–1₁ an equal division of the elliptical base as shown.
8. Mark out on each side of centre line 1–1₁ equal divisions of the circular top as shown.
9. Continue development as in former jobs.
10. Check circumferential lengths of development, top and base.
11. Check accuracy of development by testing with trammels or tape; the diagonal should be equal.

Note

Use all mean dimensions.

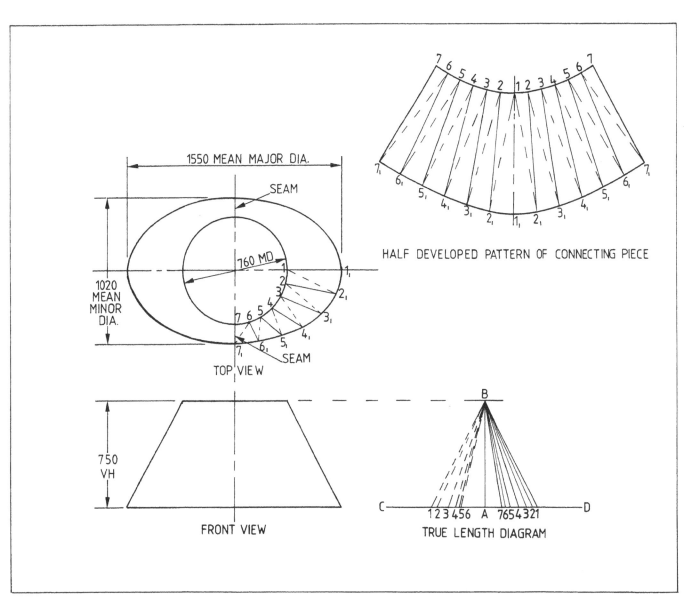

HALF DEVELOPED PATTERN OF CONNECTING PIECE

TOP VIEW

FRONT VIEW

TRUE LENGTH DIAGRAM

Development of quadrant to round transition piece by triangulation

Notes

1. Lay out using mean measurements for curved surfaces and inside measurement for square corner.

2. Check pattern for:
   (a) top curve 9–0–9 to be equal to calculate MC.
   (b) base curves 5_1–9_1 to be equal to calculated base curved length from top view
   (c) diagonals 9–9_1 to be equal.

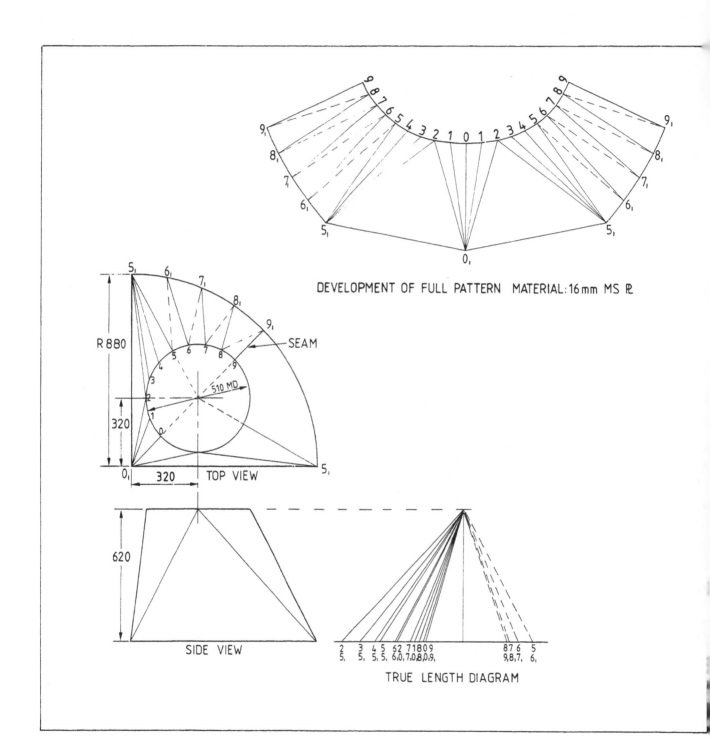

DEVELOPMENT OF FULL PATTERN   MATERIAL:16mm MS ℞

R 880

SEAM

510 MD

320

0_1 320   TOP VIEW 5_1

620

SIDE VIEW

TRUE LENGTH DIAGRAM

Exercises for developing transition pieces by triangulation

1. Develop full pattern from sketch provided of concentric round to round transition piece by triangulation (Scale 1:10).
   (a) Draw in half top view.
   (b) Draw in true length diagram.
   (c) Construct full pattern.
   (d) Calculate MD and MC of top and base.

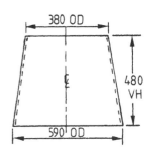

MATERIAL: 10 mm ℞

Exercises (cont'd)

2. Develop full pattern from sketch provided of offset round to round transition piece by triangulation (Scale 1:10).
   (a) Draw in half top view.
   (b) Draw in true length diagram.
   (c) Construct full pattern.
   (d) Calculate MD and MC of top and base.

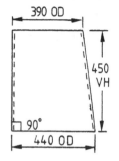

390 OD

450 VH

90°

440 OD

MATERIAL: 16 mm ℞

Exercises (cont'd)

3. Develop half pattern from sketch provided of concentric square to round transition piece by triangulation (Scale 1:10).
   (a) Draw in half top view.
   (b) Draw in true length diagram.
   (c) Construct half pattern.
   (d) Calculate MD and MC of top.

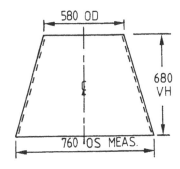

MATERIAL: 20mm ℞

4. Develop half pattern from sketch provided of the offset square to round transition piece by triangulation (Scale 1:10).
  (a) Draw in half top view.
  (b) Draw in true length diagram.
  (c) Construct half pattern.
  (d) Calculate MD and MC of top.

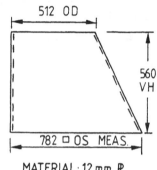

512 OD

560 VH

782 □ OS MEAS.

MATERIAL: 12 mm ℞

Exercises (cont'd)

5. Develop half pattern from the top view and front view
drawn below, showing:
(a) true length diagram;
(b) calculations.

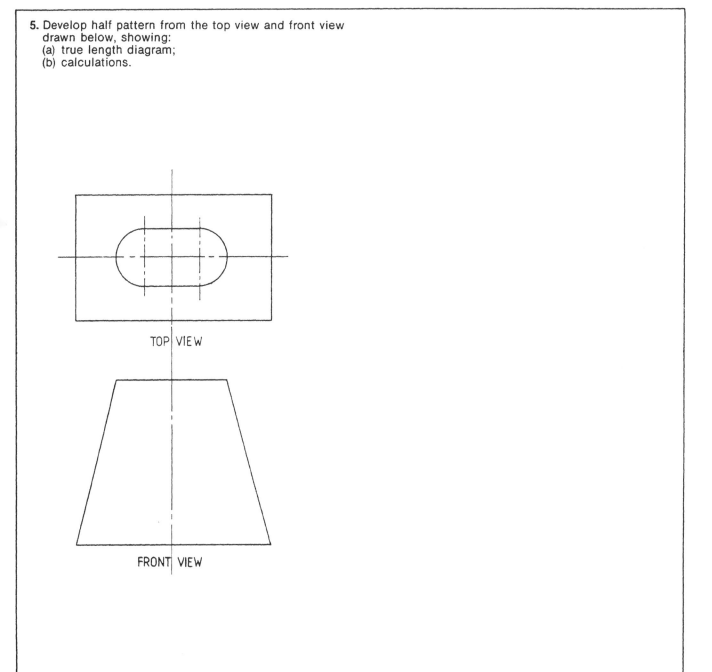

TOP VIEW

FRONT VIEW

Exercises (cont'd)

6. Develop half pattern from the top view and front view
   drawn below, showing:
   (a) true length diagram;
   (b) calculations.

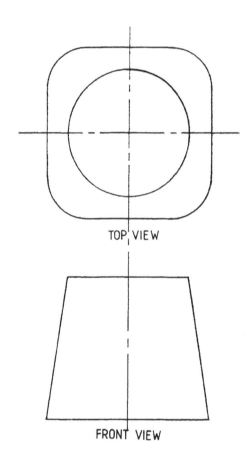

TOP VIEW

FRONT VIEW

Exercises (cont'd)

7. Develop full pattern by triangulation of the triangle to
round transition piece, from the top view and vertical
height as drawn, showing:
(a) completed true length diagram;
(b) all calculations.

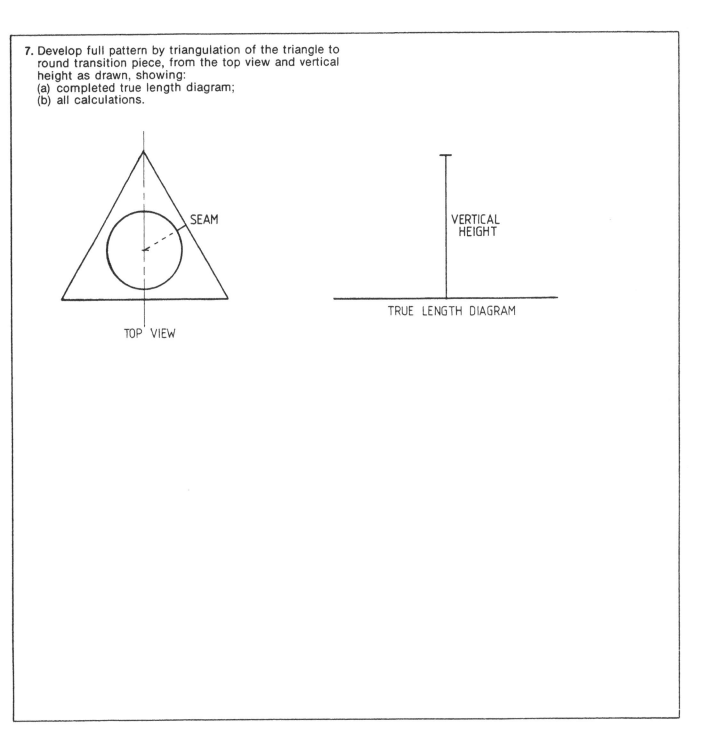

SEAM

TOP VIEW

VERTICAL
HEIGHT

TRUE LENGTH DIAGRAM

Exercises (cont'd)

8. Develop full pattern of the quadrant to round transition piece, from the top view and vertical height as drawn, showing:
(a) completed true length diagram;
(b) all calculations.

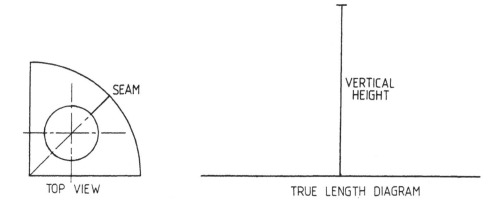

SEAM

TOP VIEW

VERTICAL
HEIGHT

TRUE LENGTH DIAGRAM

Exercises (cont'd)

9. Develop full pattern by triangulation of the ellipse to round transition piece, from the top view and vertical height as drawn, showing:
   (a) completed true length diagram;
   (b) all calculations.

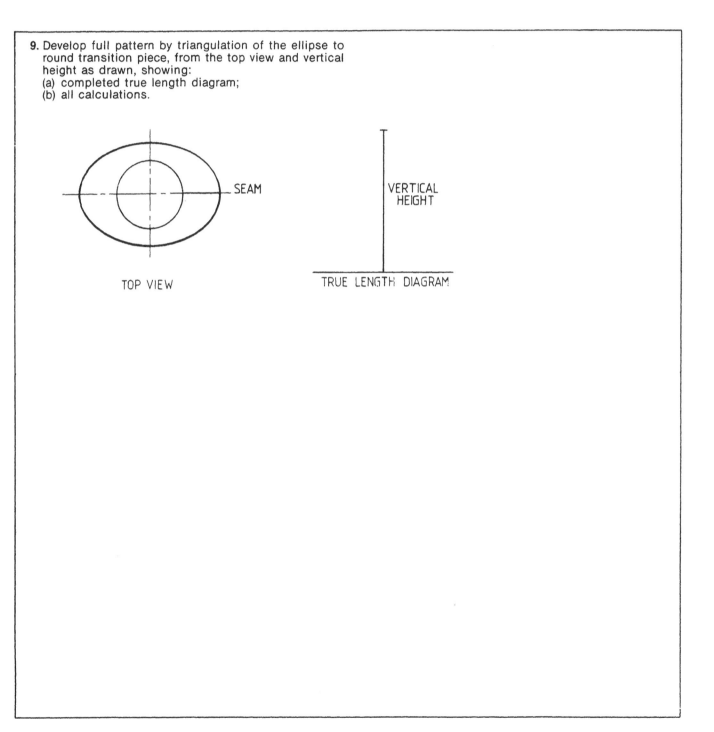

TOP VIEW

TRUE LENGTH DIAGRAM

Exercises (cont'd)

10. Develop full pattern by triangulation of the triangular shaped base with rounded corners to round top transition piece, from the top view and vertical height as drawn, showing:
(a) completed true length diagram;
(b) all calculations.

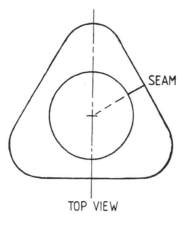

TOP VIEW

SEAM

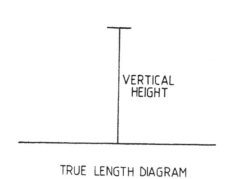

VERTICAL HEIGHT

TRUE LENGTH DIAGRAM

Part B
Triangulation of transition pieces with the ends not parallel (two-view method)

Two-view method (top or bottom view plus side view)

The following three jobs show various shapes with the end not parallel, and these jobs are developed using the same method as used in Part A of this chapter.

The earlier part of this chapter shows distinctly the use of a given top or bottom view distance placed at right angles to a common vertical height to produce a true length (slope measurement). This section (Part B) shows a similar method, except that the vertical heights are *not common* (as shown in the following sketches).

The vertical heights for the true length diagram are projected across from the front view.

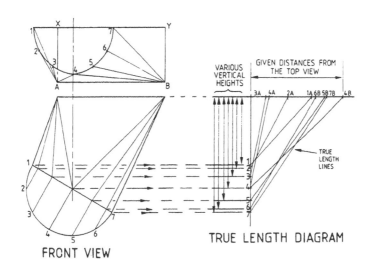

TRUE LENGTH DIAGRAM

FRONT VIEW

Following are examples for obtaining true lengths for this job:

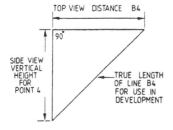

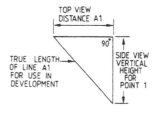

This development method is good for the shapes where the true shape of the bottom or top views are reasonably easy to draw.

With some shapes it is not economical to use this method because of the difficulty in drawing the true shape of the top or bottom views (see Part C for alternative method).

Note

Extra care must be taken to combine the correct given top view distance with the correct vertical height to create the correct true length line for use in the development.

Development of intersecting conical transition piece (top or bottom view method)

This job is developed in this part of the chapter, using given lengths from top or bottom views placed at right angles to the vertical heights from the front view to obtain the true length distances required for the development.

The variation here is that the true lengths are not all obtained from a common vertical height as in Part A of this chapter, but they are obtained from various vertical heights which are projected from given heights from the front view.

The development is similar to the develo[p]ments in Parts A and B, where you use the tr[ue] lengths of both full and broken lines and calc[u]lated divisions of both the MC for spacing (u[se] mean dimensions for both side and botto[m] views).

The development can be started from a[ny] line 1H, 4E or 7A. We have not included descri[p]tive operational steps for this part of the cha[p]ter because these appear in Part C, where t[he] same job is explained fully with progressive j[ob] steps.

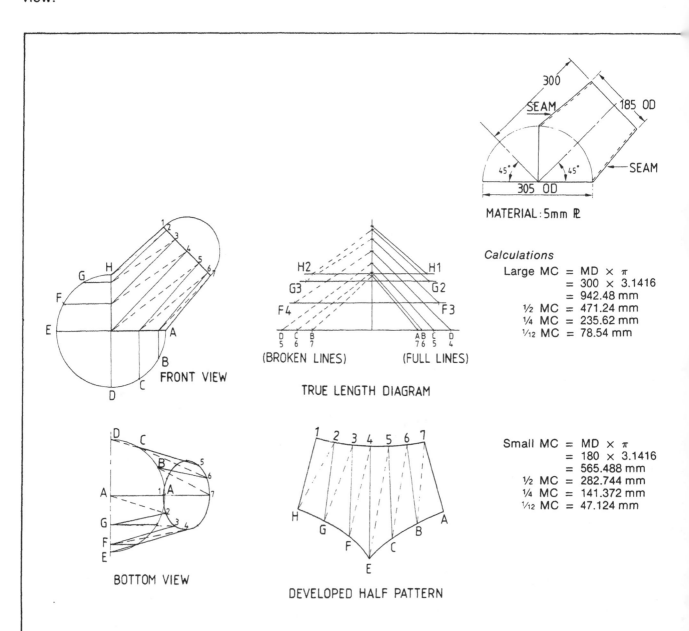

MATERIAL: 5mm ℞

Calculations

Large MC = MD × π
= 300 × 3.1416
= 942.48 mm
½ MC = 471.24 mm
¼ MC = 235.62 mm
1/12 MC = 78.54 mm

Small MC = MD × π
= 180 × 3.1416
= 565.488 mm
½ MC = 282.744 mm
¼ MC = 141.372 mm
1/12 MC = 47.124 mm

FRONT VIEW

TRUE LENGTH DIAGRAM
(BROKEN LINES) (FULL LINES)

BOTTOM VIEW

DEVELOPED HALF PATTERN

Development of square to round transition piece (ends not parallel)

Draw in both front view and top view using mean diameter for circular end and inside dimensions for square end.

This job is developed similar to Part A, using given lengths from top view placed at 90° to the various vertical heights (projected from front view) to obtain the true lengths to be used in the development.

This development must be started with the true size of triangle *A4B*. We have not included descriptive operational steps for this part of the chapter because these appear in Part C, where the same job is explained fully with progressive job steps.

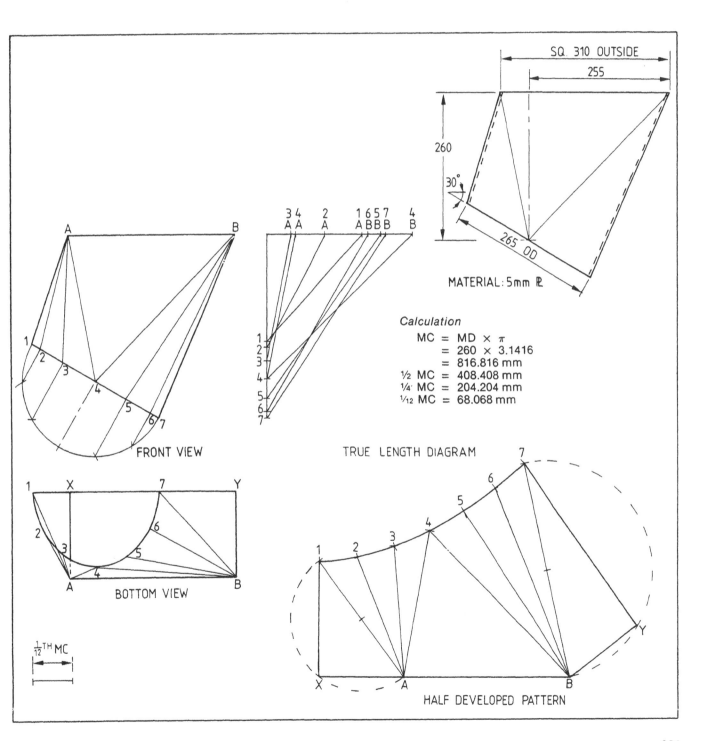

SQ. 310 OUTSIDE
255
260
30°
265 OD
MATERIAL: 5mm ℞

Calculation

$$MC = MD \times \pi$$
$$= 260 \times 3.1416$$
$$= 816.816 \text{ mm}$$
$$\tfrac{1}{2} MC = 408.408 \text{ mm}$$
$$\tfrac{1}{4} MC = 204.204 \text{ mm}$$
$$\tfrac{1}{12} MC = 68.068 \text{ mm}$$

FRONT VIEW

TRUE LENGTH DIAGRAM

BOTTOM VIEW

$\tfrac{1}{12}$ TH MC

HALF DEVELOPED PATTERN

Development of round to round transition piece (ends at 90°)

Draw side and bottom view using mean dimensions.

This job is developed similar to Part A, using given lengths from bottom view placed at 90° to the various vertical heights to obtain the tr lengths to be used in the development.

We have not included descriptive op ational job steps for this part of the chapter t cause these appear in Part C, where the sar job is explained fully with progressive j steps.

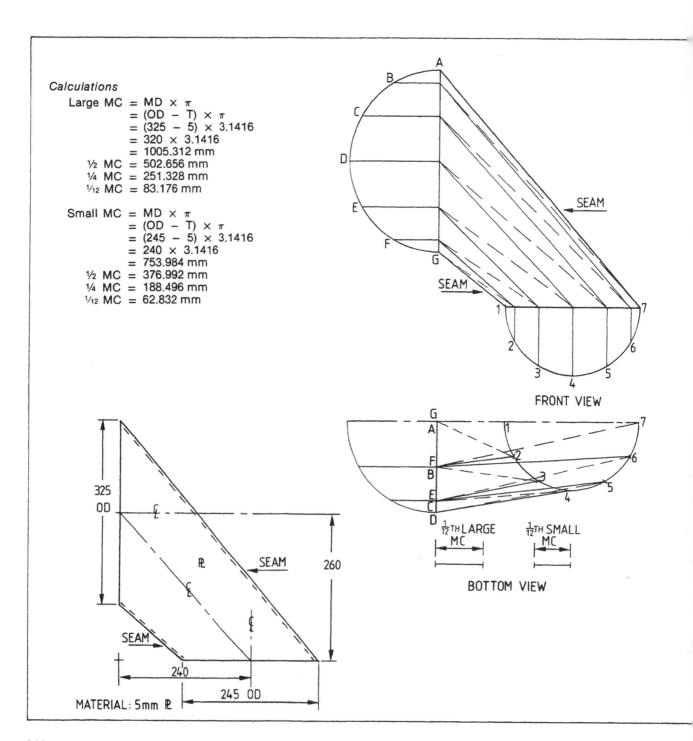

Calculations

Large MC $= MD \times \pi$
$= (OD - T) \times \pi$
$= (325 - 5) \times 3.1416$
$= 320 \times 3.1416$
$= 1005.312$ mm
½ MC $= 502.656$ mm
¼ MC $= 251.328$ mm
1/12 MC $= 83.176$ mm

Small MC $= MD \times \pi$
$= (OD - T) \times \pi$
$= (245 - 5) \times 3.1416$
$= 240 \times 3.1416$
$= 753.984$ mm
½ MC $= 376.992$ mm
¼ MC $= 188.496$ mm
1/12 MC $= 62.832$ mm

FRONT VIEW

1/12 TH LARGE MC

1/12 TH SMALL MC

BOTTOM VIEW

325 OD

260

P.

SEAM

240

245 OD

MATERIAL: 5mm P.

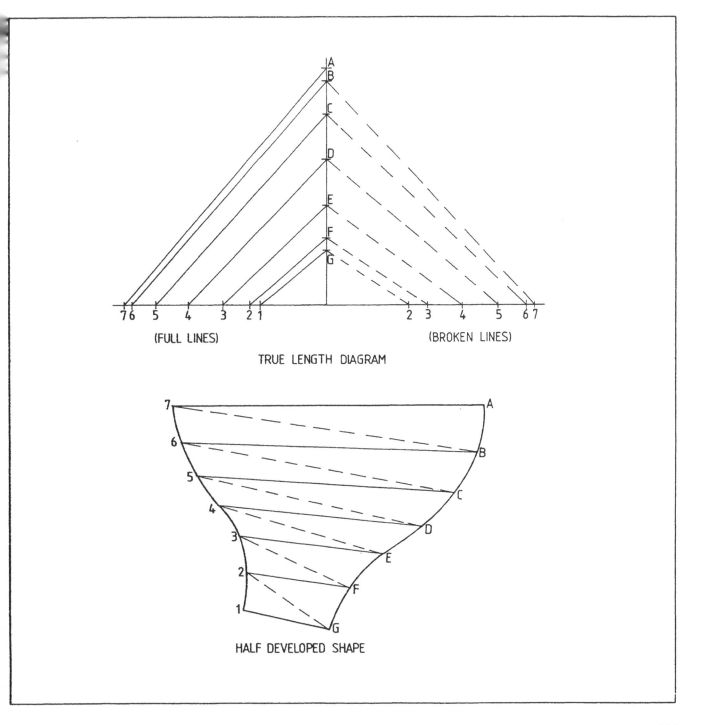

TRUE LENGTH DIAGRAM

HALF DEVELOPED SHAPE

Part C
Triangulation of transition pieces with the ends not parallel (one-view method)

One-view method (front view plus half profile of ends)

The following three jobs are the same three jobs developed in Part B, except that here they are developed using the one view (front view) method.

The front view method is when a given length from the front view is placed on a base line and the end offset distances are the heights transferred from the half profiles of the ends, then the slope line becomes the true length (see following sketches).

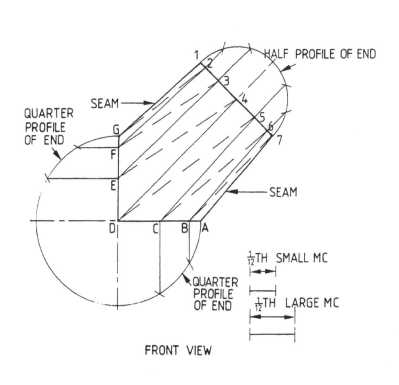

FRONT VIEW

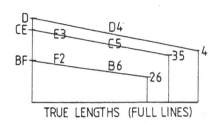

TRUE LENGTHS (FULL LINES)

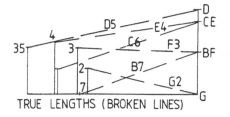

TRUE LENGTHS (BROKEN LINES)

Examples for obtaining the true lengths:

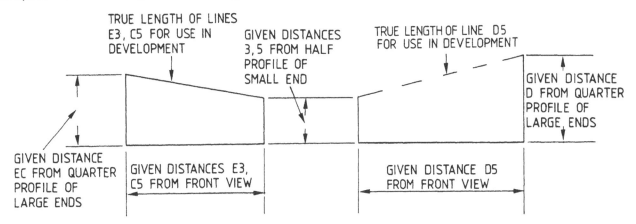

TRUE LENGTH OF LINES
E3, C5 FOR USE IN
DEVELOPMENT

GIVEN DISTANCES
3,5 FROM HALF
PROFILE OF
SMALL END

TRUE LENGTH OF LINE D5
FOR USE IN DEVELOPMENT

GIVEN DISTANCE
D FROM QUARTER
PROFILE OF
LARGE ENDS

GIVEN DISTANCE
EC FROM QUARTER
PROFILE OF
LARGE ENDS

GIVEN DISTANCES E3,
C5 FROM FRONT VIEW

GIVEN DISTANCE D5
FROM FRONT VIEW

This method (one view) is preferred for the odd shapes with the ends not parallel, because it is not necessary to construct true shapes of openings in the top or bottom views. (Only use this method on jobs that share a common centre line on the front view.)

Again, extra care must be taken to ensure that the correct length from the front view is matched up with the correct offset heights from the end half profiles, to ensure the constructed true length is correct.

Fig. 13.4 Two transition pieces forming Y piece

313

Intersecting conical transition piece (front view method)

Operation sequence

Front views, end views and true length diagram

1. Draw in front view.
2. Draw in left end view (one-quarter circle)
3. Draw in bottom view (one-quarter circle)
4. Draw in top right view (one-half circle)

   } using mean dimensions.

5. Divide left end view and bottom view into, say, three equal spaces each.
6. Divide top right view into, say, six equal spaces.
7. Draw in full development lines F2, E3, D4, C5, and B6.
8. Draw in broken development lines G2, F3, E4, D5, C6 and B7.
9. Calculate the MC of each end.
10. Mark out ¹⁄₁₂ of each MC for use in development.
11. Draw up true length diagram, using given lengths from the front view and corresponding offset distances.

Development

1. Use true length of full line D4 as starting line.
2. Using ¹⁄₁₂ of large MC as radius and point D as centre, scribe arcs both sides of point D.
3. Using ¹⁄₁₂ of small MC as radius and point 4 as centre, scribe arcs both sides of point 4.
4. Use point D as centre and true length of broken line D5 as radius to locate point 5.
5. Use point 5 as centre and true length of full line 5C as radius to locate point C.
6. Use point 4 as centre and true length of full line 4E as radius to locate point E.
7. Use point E as centre and true length of full line E3 as radius to locate point 3.
8. Using ¹⁄₁₂ of large MC as radius and points C and E as centres, scribe arcs.
9. Using ¹⁄₁₂ of large MC as radius and points 3 and 5 as centres, scribe arcs.
10. Use point C as centre and true length of broken line C6 as radius to locate point 6.
11. Use point 6 as centre and true length of full line 6B as radius to locate point B.
12. Use point 3 as centre and true length of broken line 3F as radius to locate point F.

13. Use point F as centre and true length of line F2 as radius to locate point 2.
14. Using ¹⁄₁₂ of large MC as radius and points and B as centres, scribe arcs.
15. Using ¹⁄₁₂ of small MC as radius and poi 2 and 6 as centres, scribe arcs.
16. Use point B as centre and true length broken line B7 as radius to locate point
17. Use point 7 as centre and true length of line 7A as radius to locate point A.
18. Use point 2 as centre and true length broken line 2G as radius to locate point
19. Use point G as centre and true length of line G1 as radius to locate point 1.

Checking

1. Check measurement around top curved li 1 to 7 to be equal to half MC (282.744 mm
2. Check measurement around bottom curve line G to D to be equal to one-quarter M (235.62 mm).
3. Check measurement around bottom curve line D to A to be equal to one-quarter M (235.62 mm).
4. Check diagonal 1A to be equal to diagon 7G.

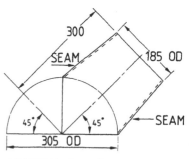

MATERIAL: 5mm ℞

Development of intersecting conical transition piece (front view method)

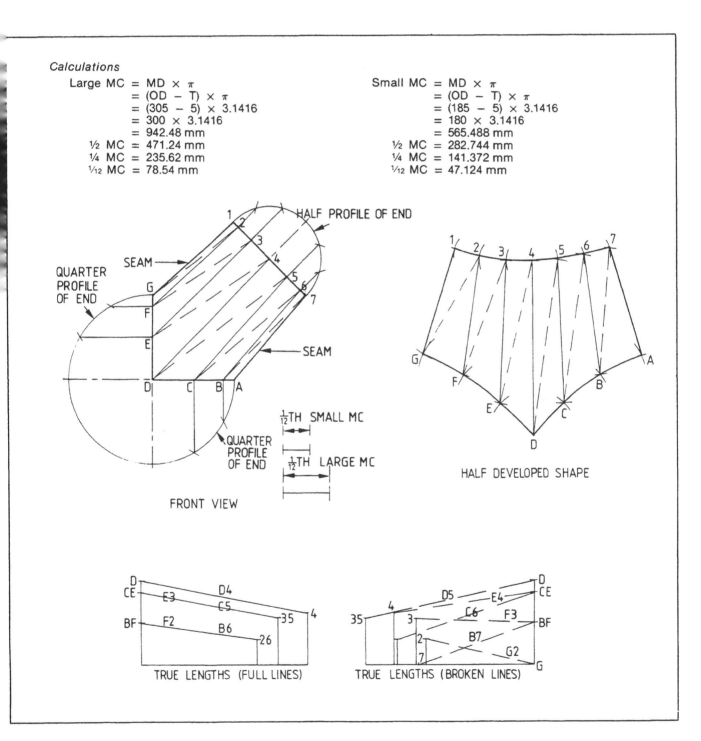

Calculations

Large MC = MD × π
= (OD − T) × π
= (305 − 5) × 3.1416
= 300 × 3.1416
= 942.48 mm
½ MC = 471.24 mm
¼ MC = 235.62 mm
¹⁄₁₂ MC = 78.54 mm

Small MC = MD × π
= (OD − T) × π
= (185 − 5) × 3.1416
= 180 × 3.1416
= 565.488 mm
½ MC = 282.744 mm
¼ MC = 141.372 mm
¹⁄₁₂ MC = 47.124 mm

HALF PROFILE OF END

SEAM

QUARTER PROFILE OF END

SEAM

QUARTER PROFILE OF END

¹⁄₁₂TH SMALL MC

¹⁄₁₂TH LARGE MC

FRONT VIEW

HALF DEVELOPED SHAPE

TRUE LENGTHS (FULL LINES)

TRUE LENGTHS (BROKEN LINES)

Square to round transition piece (ends not parallel) (front view method)

Operation sequence

Front view, end views and true length diagram

1. Draw in front view
2. Draw in half top view
3. Draw in half bottom view

} using inside dimensions for square end and mean dimensions for round end.

4. Divide half bottom view into any number of equal spaces (say six) then draw in development lines on side view.
5. Draw up true length diagram, using the given lengths from the front view and the corresponding heights from the top and bottom views, and the slope line becomes the true length.
6. Calculate the MC of the round end:

$$MC = MD \times \pi$$
$$= 260 \times 3.1416$$
$$= 816.816 \text{ mm}$$
$$\tfrac{1}{2} \text{ MC} = 408.408 \text{ mm}$$
$$\tfrac{1}{4} \text{ MC} = 204.204 \text{ mm}$$
$$\tfrac{1}{12} \text{ MC} = 68.068 \text{ mm}$$

7. Mark out $\tfrac{1}{12}$ of MC for use in the development.

Development

1. Mark front view distance *AD* on base line.
2. Using true length distance *A4* and point *A* as centre, scribe an arc.
3. Using true length distance *D4* and point *D* as centre, locate point 4.
4. Using $\tfrac{1}{12}$ MC as radius and point 4 as centre, scribe arcs both sides of point 4.
5. Using true length distance *A3* and point *A* as centre, scribe an arc to locate point 3.
6. Using true length *D5* as radius and point *D* as centre, scribe an arc to locate point 5.
7. Using $\tfrac{1}{12}$ MC as radius and points 3 and 5 as centres, scribe arcs.
8. Using true length *A2* and point *A* as centre, scribe an arc to locate point 2.
9. Using true length *D6* and point *D* as centre, scribe an arc to locate point 6.
10. Using $\tfrac{1}{12}$ MC as radius and points 2 and 6 as centres, scribe arcs.
11. Using true length *A1* and point *A* as centre, scribe an arc to locate point 1.
12. Using true length *D7* and point *D* as centre, scribe an arc to locate point 7.

13. Using distances *AB* and *DC* from half top view as radius and points *A* and *D* centres, scribe arcs.
14. Using true length *B1* from front view and point 1 as centre, scribe an arc to locate point *B*.
15. Using true length *C7* from front view and point 7 as centre, scribe an arc to locate point *C*.
16. Check angle *1BA* for 90°
17. Check angle *7CD* for 90°

} using angle in semicircle method.

18. Check running measurement around curve line 1-2-3-4-5-6-7 to be equal to half the calculated MC (817 mm).

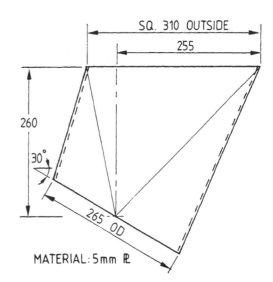

MATERIAL: 5mm ℞

316

Development of square to round transition piece (front view method)

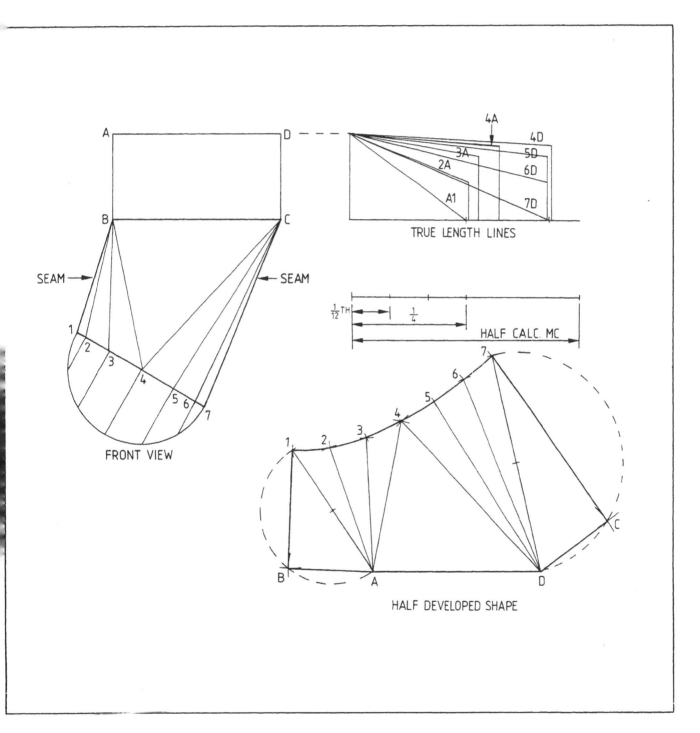

TRUE LENGTH LINES

HALF CALC. MC

FRONT VIEW

HALF DEVELOPED SHAPE

Round to round transition piece (ends at 90°) (front view method)

Operation sequence

Front view, end views and true length diagram

1. Draw in front view ⎫
2. Draw in half left side view ⎬ using mean diameters.
3. Draw in half bottom view ⎭
4. Divide half left end view and half bottom view into, say, six equal spaces each.
5. Draw in full development lines F6, E5, D4, C3 and B2 on front view.
6. Draw in broken development lines G6, F5, E4, D3, C2 and B1 on side view.
7. Draw up true length diagrams using full and broken development lines from front view and distances from end and bottom views at each end and the slope line becomes the true length.
8. Calculate the mean circumferences of each end:

$$MC = MD \times \pi$$
$$= 320 \times 3.1416$$
$$= 1005.312 \text{ mm}$$
½ MC = 502.656 mm
¼ MC = 251.328 mm
¹⁄₁₂ MC = 83.776 mm

$$MC = MD \times \pi$$
$$= 240 \times 3.1416$$
$$= 753.984 \text{ mm}$$
½ MC = 376.992 mm
¼ MC = 188.496 mm
¹⁄₁₂ MC = 62.832 mm

9. Mark out ¹⁄₁₂ of each MC for use in development

Development

1. Use line A1 from front view (true length) as start line.
2. Using ¹⁄₁₂ of large MC as radius, scribe an arc from point A.
3. Use true length of line 1B as radius and point 1 as centre, to locate point B.
4. Using ¹⁄₁₂ of small MC as radius scribe an arc from point 1.
5. Use true length of line 2B as radius and point B as centre to locate point 2.
6. Using ¹⁄₁₂ of large MC as radius, scribe an arc from point A.
7. Use true length of line 2C as radius and point 2 as centre to locate point C.

8. Using ¹⁄₁₂ of small MC as radius, scribe a arc from point 2.
9. Use true length of line 3C as radius an point C as centre to locate point 3.
10. Using ¹⁄₁₂ of large MC as radius, scribe an ar from point C.
11. Use true length of line 3D as radius an point 3 as centre to locate point D.
12. Using ¹⁄₁₂ of small MC as radius, scribe a arc from point 3.
13. Use true length of line 4D as radius an point D as centre to locate point 4.
14. Using ¹⁄₁₂ of large MC as radius, scribe an ar from point D.
15. Use true length of line 4E as radius an point 4 as centre to locate point E.
16. Using ¹⁄₁₂ of small MC as radius, scribe a arc from point 4.
17. Use true length of line E5 as radius an point E as centre to locate point 5.
18. Using ¹⁄₁₂ of large MC as radius, scribe an ar from point E.
19. Use true length of line 5F as radius an point 5 as centre to locate point F.
20. Using ¹⁄₁₂ of small MC as radius, scribe ar arc from point 5.
21. Use true length of line F6 as radius and point F as centre to locate point 6.
22. Using ¹⁄₁₂ of large MC as radius, scribe an arc from point F.
23. Use true length of line G6 as radius and point 6 as centre to locate point G.
24. Using ¹⁄₁₂ of small MC as radius, scribe ar arc from point 6.
25. Use true length of line G7 as radius and point G as centre to locate point 7.
26. Check curved line GA to be equal to ½ large MC (502.656 mm).
27. Check curved line 7–1 to be equal to ½ small MC (376.992 mm).

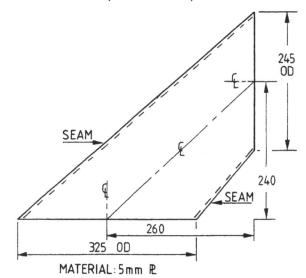

SEAM

245 OD

240

SEAM

260

325 OD

MATERIAL: 5mm ℞

318

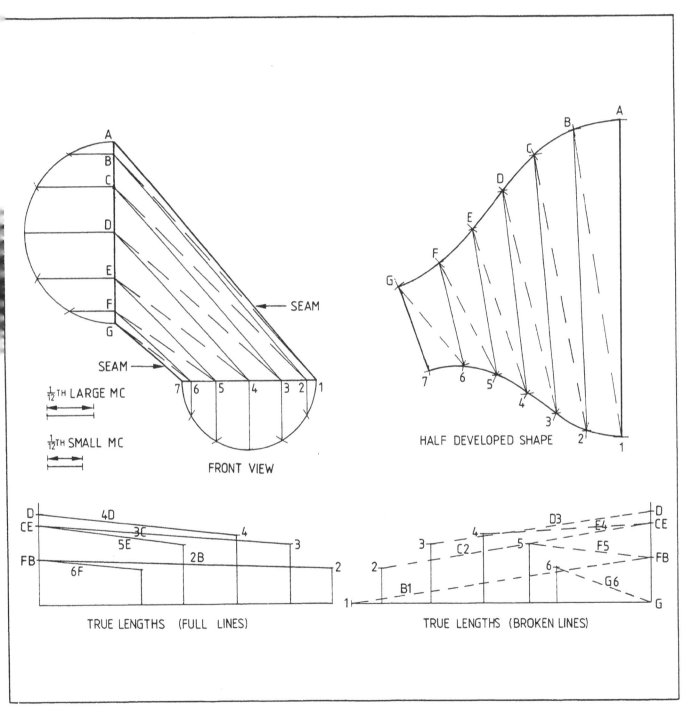

$\frac{1}{12}$TH LARGE MC

$\frac{1}{12}$TH SMALL MC

SEAM

SEAM

FRONT VIEW

HALF DEVELOPED SHAPE

TRUE LENGTHS (FULL LINES)

TRUE LENGTHS (BROKEN LINES)

Exercises for developing transition pieces using alternative method of triangulation

1. Develop half pattern from the front view
drawn below to mean dimensions, showing:
(a) true length diagrams;
(b) all calculations.

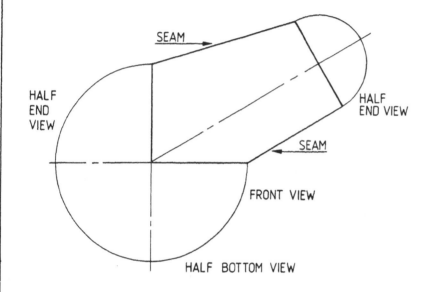

Exercises (cont'd)

2. Develop half pattern from the front view
   drawn below to mean dimensions, showing:
   (a) all calculations;
   (b) true length diagrams.

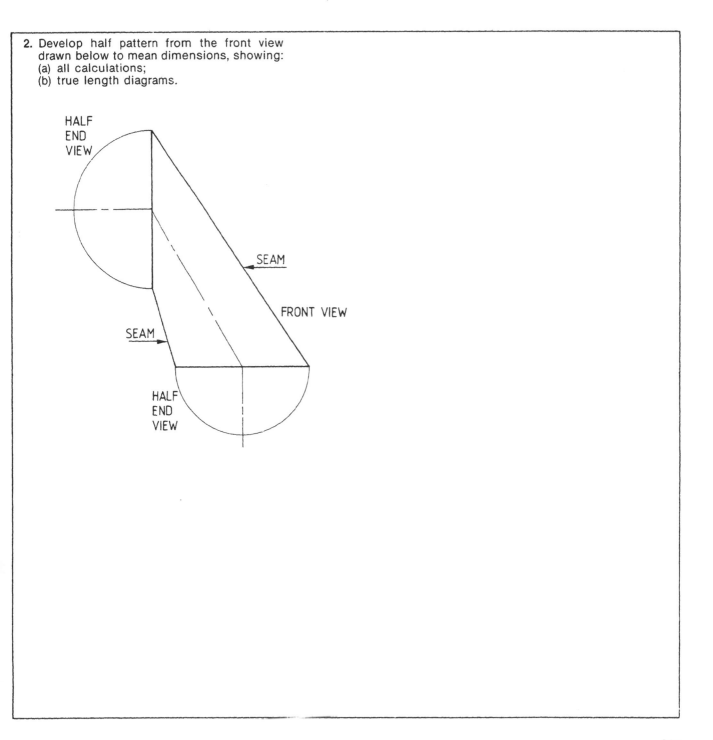

3. Develop half pattern of the transition piece drawn below with inside dimensions on the square end and mean dimensions on the circular end, showing:
   (a) all calculations;
   (b) true length diagram.

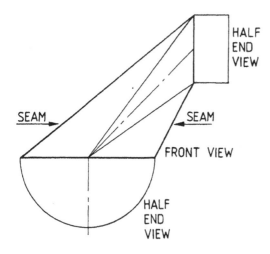

HALF END VIEW

SEAM

SEAM

FRONT VIEW

HALF END VIEW

4. Develop half pattern from the front view drawn below with inside dimensions on the rectangular end and mean dimensions on the circular end, showing:
(a) all calculations;
(b) true length diagrams.

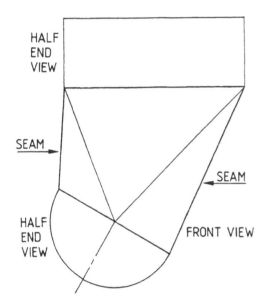

HALF
END
VIEW

SEAM

HALF
END
VIEW

SEAM

FRONT VIEW

NOTES

14

Hoppers: Square or rectangular

Fabricated hoppers are generally used as storage and/or feed bins of various application.

Hoppers, from the fabrication angle, can be categorised as those with top and base parallel or those with top and base not parallel.

Note
Do not measure from the diagrams except in exercises, because many figures are indication diagrams only and are not drawn to scale. All measurements on the diagrams are in millimetres unless stated otherwise.

Hoppers

The first group (hoppers with top and base parallel) consists of simple forms in which the shape of the development plates can be constructed without the necessity of a layout. This is done by determining the true lengths of the side plate centre lines with the aid of calculations and a knowledge of the theorem of Pythagoras, and taking remaining dimensions from the drawing.

The corner bevels can be determined from the developed plates as shown on pages 327, 329 and 331.

The hopper shown on page 327 is the simplest and most prevalent form used. All the side plates and corner bevels are the same, the corner bevels all being greater than 90°.

Page 328 shows the hopper where all the developed plates and corner bevels are different, but all corner bevels are greater than 90°.

Page 330 indicates the type of hopper which, while still similar to that on page 328, has two corner bevels greater than 90° and the remaining two corner bevels less than 90°. The method of finding the corner bevels less than 90° is as shown.

Hoppers as shown on page 332 are more complex than those previously dealt with, and the layout of top view, end view and front view as indicated can simplify the construction of developed plates.

Plates A and C can be constructed as for the other hoppers, whereas plate B will require the application of triangulation techniques.

The break or setting line will break the side plates into the required triangles as in top view.

The true lengths can be determined as illustrated. Corner bevels can be determined as for previous hoppers. To find the amount of setting angle for the side plate B, work from the developed plates and end view as indicated on the sketch.

Remember: When marking out developed plates for any hopper, always check the matching corner lengths, which must be equal.

In this chapter we are not giving a simplified breakdown of steps because the previous chapter shows how to develop true length using offset distance from top view at right angles to vertical height. This true length is then used as the length of the centre line of the developed plate.

Pages 334–5 show the development of the leg of the saw horse (simplest method). Whilst this job is not a hopper, the legs are similar to the corner of a regular hopper and the pressing angle for the leg is determined as on pages 327, 329 and 331.

Regular hopper: Frustum of pyramid

No layout actually required. True length of the developed plate of the regular hopper can be calculated using Pythagoras' Theorem shown below:

$$S = \sqrt{\left(\frac{A - B}{2}\right)^2 + P^2}$$

When checking the development of the plate the corner length of the regular hopper can be checked using a similar calculation:

$$\text{Corner length} = \sqrt{\left(\frac{A - B}{2}\right)^2 + P^2}$$

where P is equal to answer from the first calculation.

Note The four side plates are all equal in size and the four corner bevels are equal.

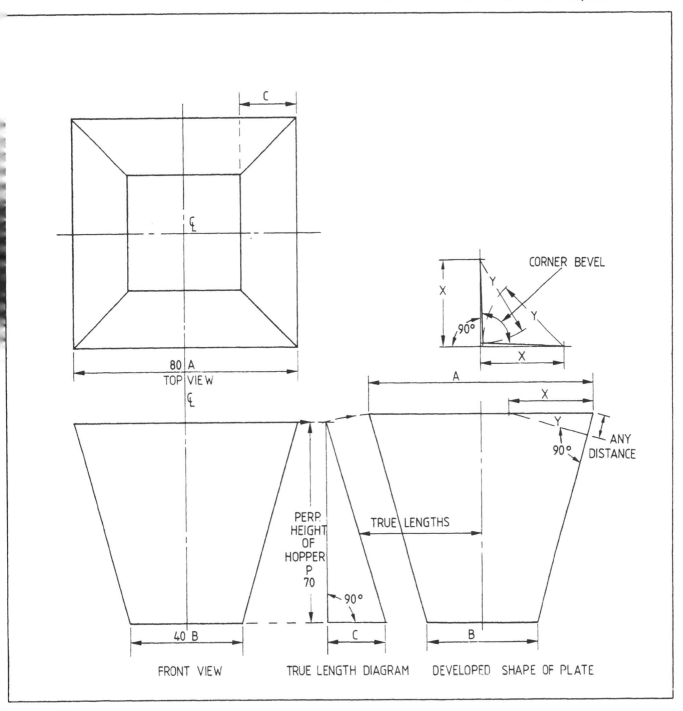

Hopper, top and base parallel, off centre, all corner bevels are greater than 90°

Development using top view, front view and true lengths. True lengths developed using offset distances from top view and vertical height from front view.

All developed plates are different in size. All corner bevels are different. Develop the other three matching corner bevels using similar method as shown below.

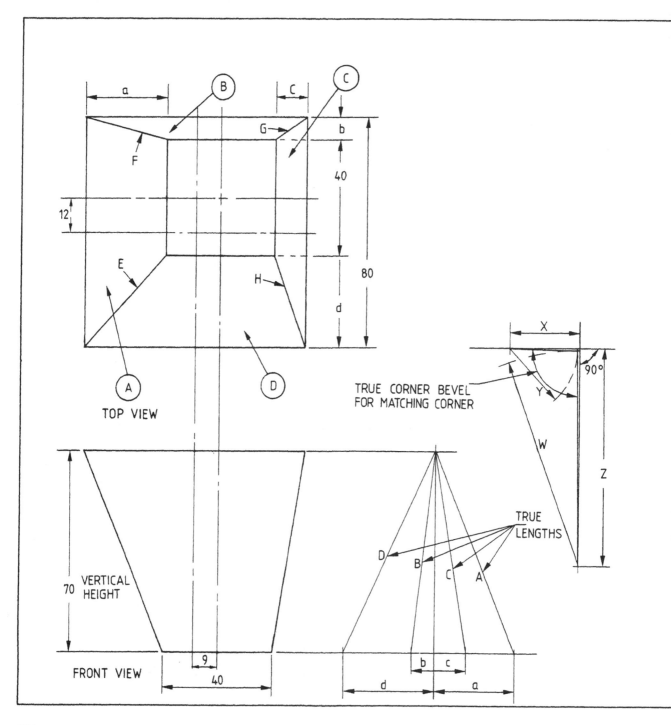

TOP VIEW

FRONT VIEW

VERTICAL HEIGHT

TRUE CORNER BEVEL FOR MATCHING CORNER

TRUE LENGTHS

Development of hopper (cont'd)

The four developed plates, one required of each

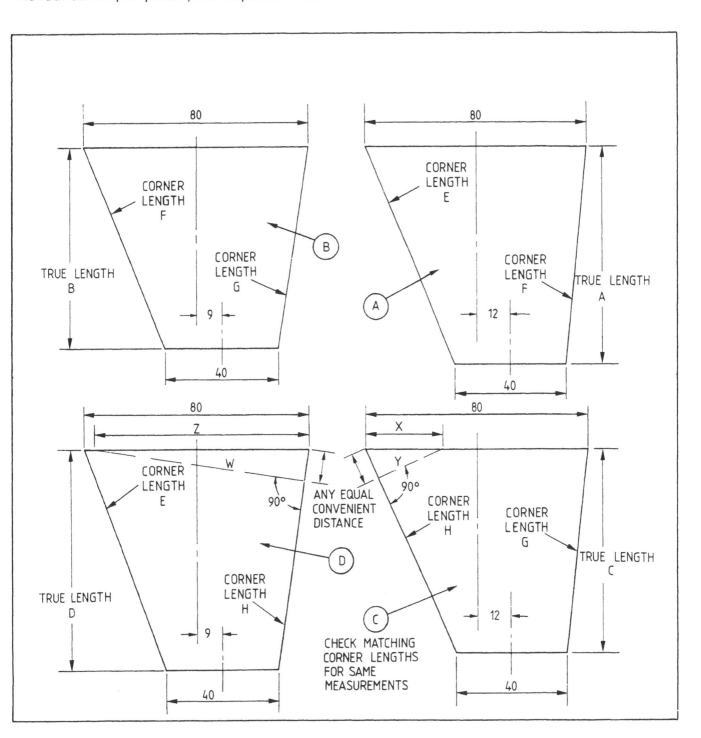

Formulae applicable to development by calculation

True length of centre line of developed side plates for eccentric hoppers can be calculated using Pythagoras' Theorem shown below:

$$TLB = \sqrt{b^2 + VH^2}$$

When checking the development of the plate the corner lengths of eccentric hoppers can be checked using a similar calculation:

$$CLF = \sqrt{a^2 + b^2 + VH^2}$$

Hopper, top and base parallel and offset, two open and two closed corner bevels

Development for one of the closed corner bevels. Develop the other closed bevel similarly. Develop the two open bevels as in the past exercises.

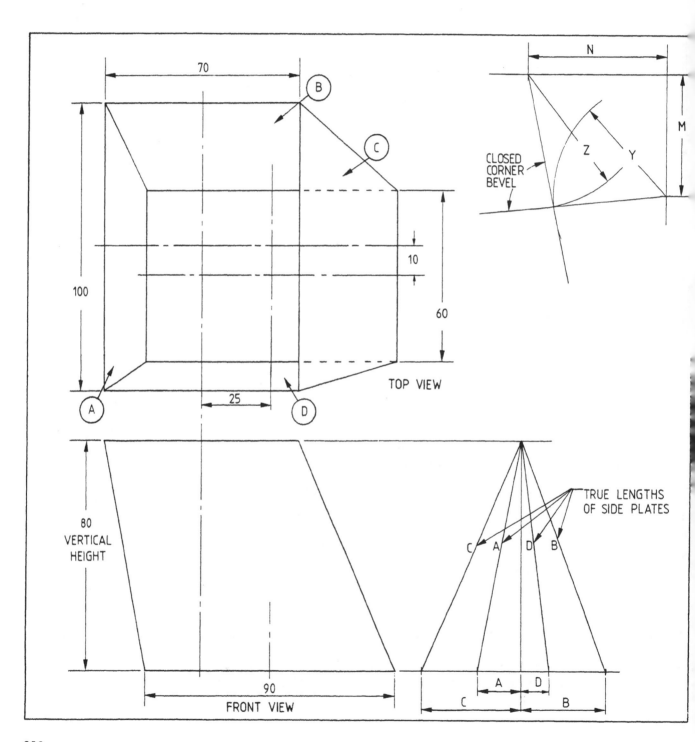

Development of hopper (cont'd)

The four developed plates, one required of each

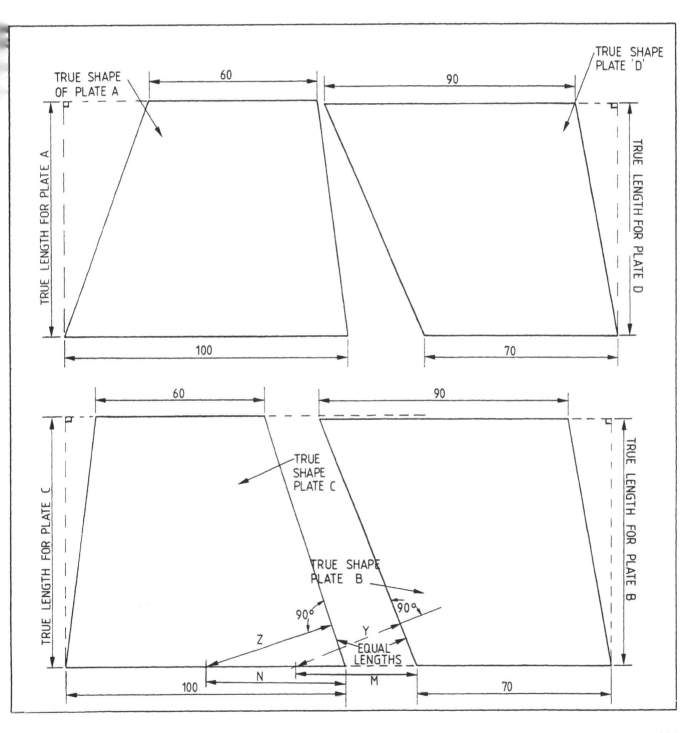

Hopper, top and base not parallel, four open corner bevels

Development of the four open corner bevels will be the same as on pages 327, 329 and 331.

Right end view required in this job because the bottom is not parallel to the top. Offset distance *X* required to develop the setting angle for the side plates item *B*.

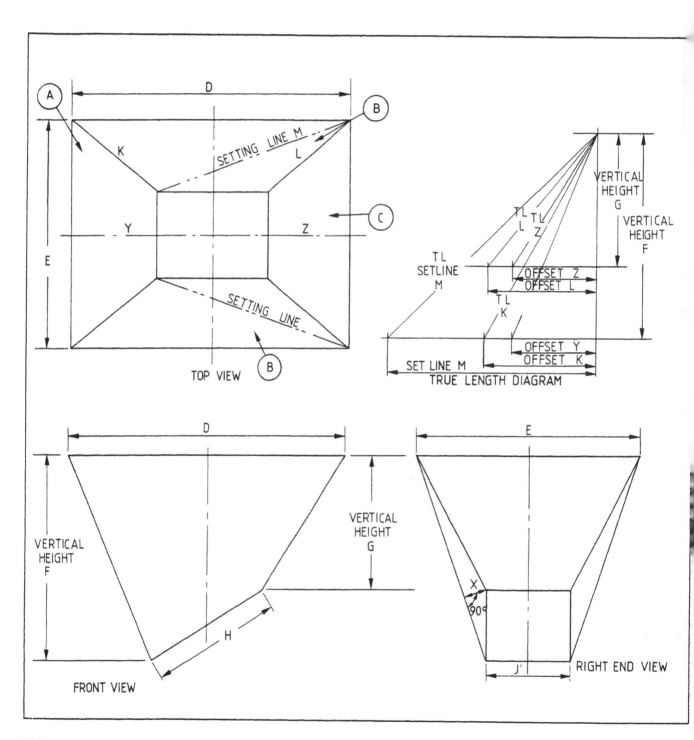

Development of hopper (cont'd)

Developed shapes of plates and true length diagram

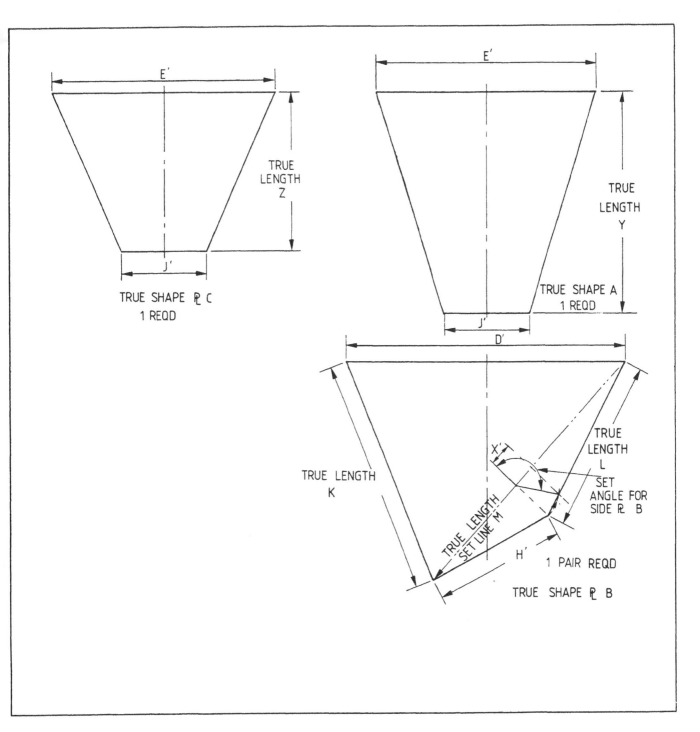

E'

TRUE
LENGTH
Z

J'

TRUE SHAPE ℞ C
1 REQD

E'

TRUE
LENGTH
Y

J'

TRUE SHAPE A
1 REQD

D'

TRUE LENGTH
K

X'

TRUE
LENGTH
L

SET
ANGLE FOR
SIDE ℞ B

TRUE LENGTH
SET LINE M

H'

1 PAIR REQD

TRUE SHAPE ℞ B

Laying out of saw horse leg (simplest way)

To lay out the pattern of one leg of the saw horse, we use the true shape of the flat surfaces A and B. These are obtained by looking at the surfaces in the direction of arrows A and B, or alternatively tilting the saw horse towards you until each surface in turn is vertical and 90° to your vision. By following the marking steps 1 to 6, the result will be one leg marked to shape.

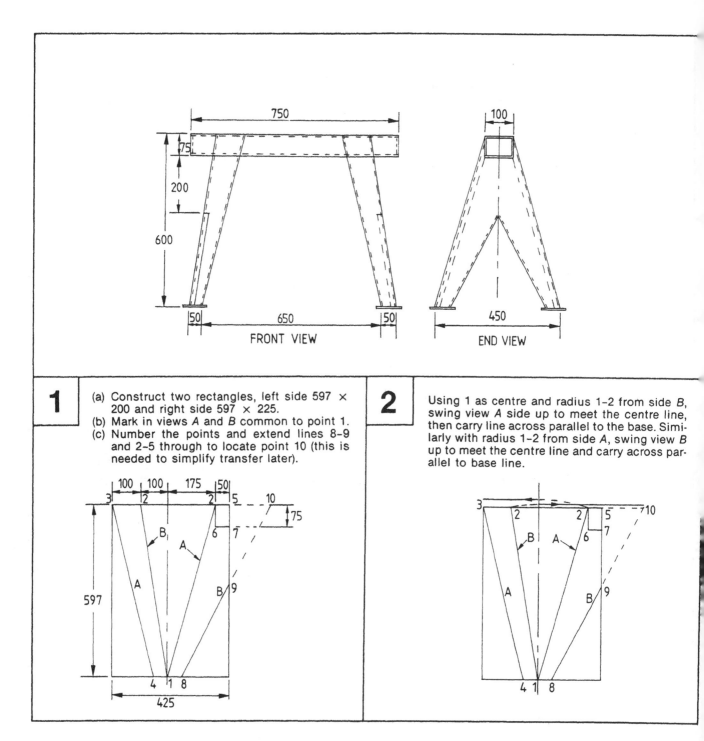

FRONT VIEW

END VIEW

1
(a) Construct two rectangles, left side 597 × 200 and right side 597 × 225.
(b) Mark in views A and B common to point 1.
(c) Number the points and extend lines 8–9 and 2–5 through to locate point 10 (this is needed to simplify transfer later).

2
Using 1 as centre and radius 1–2 from side B, swing view A side up to meet the centre line, then carry line across parallel to the base. Similarly with radius 1–2 from side A, swing view B up to meet the centre line and carry across parallel to base line.

3	Extend the points 3–2–10 vertical to locate the true length points 3′–2′–10′. Mark in the new positions of points 6′–5′–7′–9′. We now have the true length shape of the sides *A* and *B*. (Check line 1′–2′ in both views *A* and *B* is same length.)

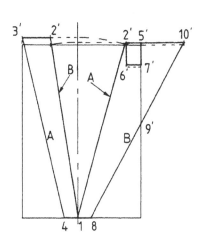

4	Lay out on template plate *A* centre line equal in length to line 1′–2′ from step 3. Transfer the true shape of sides *A* and *B* to the template marked around centre line 1′–2′ (transfer shapes from step 3) to locate points 3–4–8–10.

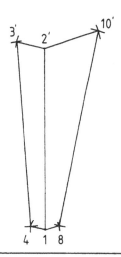

5	Then locate points 5–6–7–9 on step 5 transferred from true length step 3.

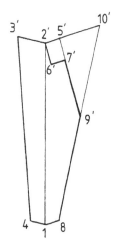

6	Now add the flanges to the edges 3–4 and 8–9. The angles for the ends of the flanges are transferred as shown below (step 6) as *a*, *b* and *c*. This template is now ready for marking as two off as drawn and two off opposite hand. Now ready for bending, refer to pages 327, 329 and 331 for obtaining true pressing angle.

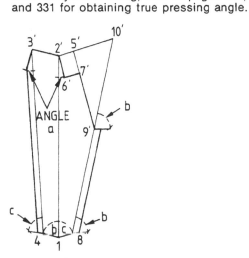

335

Fig. 14.1 Checking the accuracy of the completed saw horse